FREE Test Taking Tips DVD Offer

To help us better serve you, we have developed a Test Taking Tips DVD that we would like to give you for FREE. **This DVD covers world-class test taking tips that you can use to be even more successful when you are taking your test.**

All that we ask is that you email us your feedback about your study guide. Please let us know what you thought about it – whether that is good, bad or indifferent.

To get your **FREE Test Taking Tips DVD**, email freedvd@studyguideteam.com with "FREE DVD" in the subject line and the following information in the body of the email:

   a. The title of your study guide.

   b. Your product rating on a scale of 1-5, with 5 being the highest rating.

   c. Your feedback about the study guide. What did you think of it?

   d. Your full name and shipping address to send your free DVD.

If you have any questions or concerns, please don't hesitate to contact us at freedvd@studyguideteam.com.

Thanks again!

ACT Math Prep Book 2018 & 2019

ACT Math Workbook & Practice Tests for the ACT Exam

Test Prep Books College Entrance Study Guide Review Team

Table of Contents

Quick Overview

As you draw closer to taking your exam, effective preparation becomes more and more important. Thankfully, you have this study guide to help you get ready. Use this guide to help keep your studying on track and refer to it often.

This study guide contains several key sections that will help you be successful on your exam. The guide contains tips for what you should do the night before and the day of the test. Also included are test-taking tips. Knowing the right information is not always enough. Many well-prepared test takers struggle with exams. These tips will help equip you to accurately read, assess, and answer test questions.

A large part of the guide is devoted to showing you what content to expect on the exam and to helping you better understand that content. In this guide are practice test questions so that you can see how well you have grasped the content. Then, answer explanations are provided so that you can understand why you missed certain questions.

Don't try to cram the night before you take your exam. This is not a wise strategy for a few reasons. First, your retention of the information will be low. Your time would be better used by reviewing information you already know rather than trying to learn a lot of new information. Second, you will likely become stressed as you try to gain a large amount of knowledge in a short amount of time. Third, you will be depriving yourself of sleep. So be sure to go to bed at a reasonable time the night before. Being well-rested helps you focus and remain calm.

Be sure to eat a substantial breakfast the morning of the exam. If you are taking the exam in the afternoon, be sure to have a good lunch as well. Being hungry is distracting and can make it difficult to focus. You have hopefully spent lots of time preparing for the exam. Don't let an empty stomach get in the way of success!

When travelling to the testing center, leave earlier than needed. That way, you have a buffer in case you experience any delays. This will help you remain calm and will keep you from missing your appointment time at the testing center.

Be sure to pace yourself during the exam. Don't try to rush through the exam. There is no need to risk performing poorly on the exam just so you can leave the testing center early. Allow yourself to use all of the allotted time if needed.

Remain positive while taking the exam even if you feel like you are performing poorly. Thinking about the content you should have mastered will not help you perform better on the exam.

Once the exam is complete, take some time to relax. Even if you feel that you need to take the exam again, you will be well served by some down time before you begin studying again. It's often easier to convince yourself to study if you know that it will come with a reward!

Test-Taking Strategies

1. Predicting the Answer

When you feel confident in your preparation for a multiple-choice test, try predicting the answer before reading the answer choices. This is especially useful on questions that test objective factual knowledge. By predicting the answer before reading the available choices, you eliminate the possibility that you will be distracted or led astray by an incorrect answer choice. You will feel more confident in your selection if you read the question, predict the answer, and then find your prediction among the answer choices. After using this strategy, be sure to still read all of the answer choices carefully and completely. If you feel unprepared, you should not attempt to predict the answers. This would be a waste of time and an opportunity for your mind to wander in the wrong direction.

2. Reading the Whole Question

Too often, test takers scan a multiple-choice question, recognize a few familiar words, and immediately jump to the answer choices. Test authors are aware of this common impatience, and they will sometimes prey upon it. For instance, a test author might subtly turn the question into a negative, or he or she might redirect the focus of the question right at the end. The only way to avoid falling into these traps is to read the entirety of the question carefully before reading the answer choices.

3. Looking for Wrong Answers

Long and complicated multiple-choice questions can be intimidating. One way to simplify a difficult multiple-choice question is to eliminate all of the answer choices that are clearly wrong. In most sets of answers, there will be at least one selection that can be dismissed right away. If the test is administered on paper, the test taker could draw a line through it to indicate that it may be ignored; otherwise, the test taker will have to perform this operation mentally or on scratch paper. In either case, once the obviously incorrect answers have been eliminated, the remaining choices may be considered. Sometimes identifying the clearly wrong answers will give the test taker some information about the correct answer. For instance, if one of the remaining answer choices is a direct opposite of one of the eliminated answer choices, it may well be the correct answer. The opposite of obviously wrong is obviously right! Of course, this is not always the case. Some answers are obviously incorrect simply because they are irrelevant to the question being asked. Still, identifying and eliminating some incorrect answer choices is a good way to simplify a multiple-choice question.

4. Don't Overanalyze

Anxious test takers often overanalyze questions. When you are nervous, your brain will often run wild, causing you to make associations and discover clues that don't actually exist. If you feel that this may be a problem for you, do whatever you can to slow down during the test. Try taking a deep breath or counting to ten. As you read and consider the question, restrict yourself to the particular words used by the author. Avoid thought tangents about what the author *really* meant, or what he or she was *trying* to say. The only things that matter on a multiple-choice test are the words that are actually in the question. You must avoid reading too much into a multiple-choice question, or supposing that the writer meant something other than what he or she wrote.

5. No Need for Panic

It is wise to learn as many strategies as possible before taking a multiple-choice test, but it is likely that you will come across a few questions for which you simply don't know the answer. In this situation, avoid panicking. Because most multiple-choice tests include dozens of questions, the relative value of a single wrong answer is small. As much as possible, you should compartmentalize each question on a multiple-choice test. In other words, you should not allow your feelings about one question to affect your success on the others. When you find a question that you either don't understand or don't know how to answer, just take a deep breath and do your best. Read the entire question slowly and carefully. Try rephrasing the question a couple of different ways. Then, read all of the answer choices carefully. After eliminating obviously wrong answers, make a selection and move on to the next question.

6. Confusing Answer Choices

When working on a difficult multiple-choice question, there may be a tendency to focus on the answer choices that are the easiest to understand. Many people, whether consciously or not, gravitate to the answer choices that require the least concentration, knowledge, and memory. This is a mistake. When you come across an answer choice that is confusing, you should give it extra attention. A question might be confusing because you do not know the subject matter to which it refers. If this is the case, don't eliminate the answer before you have affirmatively settled on another. When you come across an answer choice of this type, set it aside as you look at the remaining choices. If you can confidently assert that one of the other choices is correct, you can leave the confusing answer aside. Otherwise, you will need to take a moment to try to better understand the confusing answer choice. Rephrasing is one way to tease out the sense of a confusing answer choice.

7. Your First Instinct

Many people struggle with multiple-choice tests because they overthink the questions. If you have studied sufficiently for the test, you should be prepared to trust your first instinct once you have carefully and completely read the question and all of the answer choices. There is a great deal of research suggesting that the mind can come to the correct conclusion very quickly once it has obtained all of the relevant information. At times, it may seem to you as if your intuition is working faster even than your reasoning mind. This may in fact be true. The knowledge you obtain while studying may be retrieved from your subconscious before you have a chance to work out the associations that support it. Verify your instinct by working out the reasons that it should be trusted.

8. Key Words

Many test takers struggle with multiple-choice questions because they have poor reading comprehension skills. Quickly reading and understanding a multiple-choice question requires a mixture of skill and experience. To help with this, try jotting down a few key words and phrases on a piece of scrap paper. Doing this concentrates the process of reading and forces the mind to weigh the relative importance of the question's parts. In selecting words and phrases to write down, the test taker thinks about the question more deeply and carefully. This is especially true for multiple-choice questions that are preceded by a long prompt.

9. Subtle Negatives

One of the oldest tricks in the multiple-choice test writer's book is to subtly reverse the meaning of a question with a word like *not* or *except*. If you are not paying attention to each word in the question, you can easily be led astray by this trick. For instance, a common question format is, "Which of the following is…?" Obviously, if the question instead is, "Which of the following is not…?," then the answer will be quite different. Even worse, the test makers are aware of the potential for this mistake and will include one answer choice that would be correct if the question were not negated or reversed. A test taker who misses the reversal will find what he or she believes to be a correct answer and will be so confident that he or she will fail to reread the question and discover the original error. The only way to avoid this is to practice a wide variety of multiple-choice questions and to pay close attention to each and every word.

10. Reading Every Answer Choice

It may seem obvious, but you should always read every one of the answer choices! Too many test takers fall into the habit of scanning the question and assuming that they understand the question because they recognize a few key words. From there, they pick the first answer choice that answers the question they believe they have read. Test takers who read all of the answer choices might discover that one of the latter answer choices is actually *more* correct. Moreover, reading all of the answer choices can remind you of facts related to the question that can help you arrive at the correct answer. Sometimes, a misstatement or incorrect detail in one of the latter answer choices will trigger your memory of the subject and will enable you to find the right answer. Failing to read all of the answer choices is like not reading all of the items on a restaurant menu: you might miss out on the perfect choice.

11. Spot the Hedges

One of the keys to success on multiple-choice tests is paying close attention to every word. This is never truer than with words like almost, most, some, and sometimes. These words are called "hedges" because they indicate that a statement is not totally true or not true in every place and time. An absolute statement will contain no hedges, but in many subjects, the answers are not always straightforward or absolute. There are always exceptions to the rules in these subjects. For this reason, you should favor those multiple-choice questions that contain hedging language. The presence of qualifying words indicates that the author is taking special care with his or her words, which is certainly important when composing the right answer. After all, there are many ways to be wrong, but there is only one way to be right! For this reason, it is wise to avoid answers that are absolute when taking a multiple-choice test. An absolute answer is one that says things are either all one way or all another. They often include words like *every*, *always*, *best*, and *never*. If you are taking a multiple-choice test in a subject that doesn't lend itself to absolute answers, be on your guard if you see any of these words.

12. Long Answers

In many subject areas, the answers are not simple. As already mentioned, the right answer often requires hedges. Another common feature of the answers to a complex or subjective question are qualifying clauses, which are groups of words that subtly modify the meaning of the sentence. If the question or answer choice describes a rule to which there are exceptions or the subject matter is complicated, ambiguous, or confusing, the correct answer will require many words in order to be expressed clearly and accurately. In essence, you should not be deterred by answer choices that seem excessively long. Oftentimes, the author of the text will not be able to write the correct answer without offering some qualifications and modifications. Your job is to read the answer choices thoroughly and completely and to select the one that most accurately and precisely answers the question.

13. Restating to Understand

Sometimes, a question on a multiple-choice test is difficult not because of what it asks but because of how it is written. If this is the case, restate the question or answer choice in different words. This process serves a couple of important purposes. First, it forces you to concentrate on the core of the question. In order to rephrase the question accurately, you have to understand it well. Rephrasing the question will concentrate your mind on the key words and ideas. Second, it will present the information to your mind in a fresh way. This process may trigger your memory and render some useful scrap of information picked up while studying.

14. True Statements

Sometimes an answer choice will be true in itself, but it does not answer the question. This is one of the main reasons why it is essential to read the question carefully and completely before proceeding to the answer choices. Too often, test takers skip ahead to the answer choices and look for true statements. Having found one of these, they are content to select it without reference to the question above. Obviously, this provides an easy way for test makers to play tricks. The savvy test taker will always read the entire question before turning to the answer choices. Then, having settled on a correct answer choice, he or she will refer to the original question and ensure that the selected answer is relevant. The mistake of choosing a correct-but-irrelevant answer choice is especially common on questions related to specific pieces of objective knowledge. A prepared test taker will have a wealth of factual knowledge at his or her disposal, and should not be careless in its application.

15. No Patterns

One of the more dangerous ideas that circulates about multiple-choice tests is that the correct answers tend to fall into patterns. These erroneous ideas range from a belief that B and C are the most common right answers, to the idea that an unprepared test-taker should answer "A-B-A-C-A-D-A-B-A." It cannot be emphasized enough that pattern-seeking of this type is exactly the WRONG way to approach a multiple-choice test. To begin with, it is highly unlikely that the test maker will plot the correct answers according to some predetermined pattern. The questions are scrambled and delivered in a random order. Furthermore, even if the test maker was following a pattern in the assignation of correct answers, there is no reason why the test taker would know which pattern he or she was using. Any attempt to discern a pattern in the answer choices is a waste of time and a distraction from the real work of taking the test. A test taker would be much better served by extra preparation before the test than by reliance on a pattern in the answers.

FREE DVD OFFER

Don't forget that doing well on your exam includes both understanding the test content and understanding how to use what you know to do well on the test. We offer a completely FREE Test Taking Tips DVD that covers world class test taking tips that you can use to be even more successful when you are taking your test.

All that we ask is that you email us your feedback about your study guide. To get your **FREE Test Taking Tips DVD**, email freedvd@studyguideteam.com with "FREE DVD" in the subject line and the following information in the body of the email:

- The title of your study guide.
- Your product rating on a scale of 1-5, with 5 being the highest rating.
- Your feedback about the study guide. What did you think of it?
- Your full name and shipping address to send your free DVD.

Introduction to the ACT

Function of the Test

The ACT is one of two national standardized college entrance examinations (the SAT being the other). Most prospective college students take the ACT or the SAT, and it is increasingly common for students to take both. All four-year colleges and universities in the United States accept the ACT for admissions purposes, and some require it. Some of those schools also use ACT subject scores for placement purposes. Sixteen states also require all high school juniors to take the ACT as part of the states' school evaluation efforts.

The vast majority of people taking the ACT are high school juniors and seniors who intend to apply to college. Traditionally, the SAT was more commonly taken than the ACT, particularly among students on the East and West coasts. However, the popularity of the ACT has grown dramatically in recent years and is now commonly taken by students in all fifty states. In fact, starting in 2013, more test takers took the ACT than the SAT. In 2015, 1.92 million students took the ACT. About 28 percent of 2015 high school graduates taking the ACT met the test's college-readiness benchmarks in all four subjects, while 31 percent met none of the benchmarks.

Test Administration

The ACT is offered on six dates throughout the year in the U.S. and Canada, and on five of those same dates in other countries. The registration fee includes score reports for four colleges, with additional reports available for purchase. There is a separate registration fee for the optional writing section.

On test dates, the ACT is administered at test centers throughout the world. The test centers are usually high schools or colleges, with several locations usually available in significant population centers.

Test takers can retake the ACT as frequently as the test is offered, up to a maximum of twelve times; although, individual colleges may have limits on how many retakes they will consider. Scores from the various sections cannot be combined from different sessions. The ACT does provide reasonable accommodations to test takers with professionally-documented disabilities.

Test Format

The ACT consists of 215 multiple-choice questions in four subject areas (English, mathematics, reading, and science) and takes about three hours and thirty minutes to complete. It also has an optional writing test, which takes an additional forty minutes.

The English section is 45 minutes long and contains 75 questions on usage, language mechanics, and rhetorical skills. The Mathematics section is 60 minutes long and contains 60 questions on algebra, geometry, and elementary trigonometry. Calculators that meet the ACT's calculator policy are permitted on the Mathematics section. The Reading section is 35 minutes long and contains four written passages with ten questions per passage. The Science section is 35 minutes long and contains 40 questions.

The Writing section is forty minutes long and is always given at the end so that test takers not wishing to take it may leave after completing the other four sections. This section consists of one essay in which students must analyze three different perspectives on a broad social issue. Although the Writing section is optional, some colleges do require it.

Section	Length	Questions
English	45 minutes	75
Mathematics	60 minutes	60
Reading	35 minutes	40
Science	35 minutes	40
Writing (optional)	40 minutes	1 essay

Scoring

Test takers receive a score between 1 and 36 for each of the four subject areas. Those scores are averaged together to give a Composite Score, which is the primary score reported as an "ACT score." The most prestigious schools typically admit students with Composite ACT Scores in the low 30's. Other selective schools typically admit students with scores in the high 20's. Traditional colleges more likely admit students with scores in the low 20's, while community colleges and other more open schools typically accept students with scores in the high teens. In 2015, the average composite score among all test takers (including those not applying to college) was 21.

Recent/Future Developments

In 2015, the Writing section underwent several changes. The allotted time extended 10 minutes (from 30 to 40 minutes) and the scoring changed to a scale from 1 to 36 (as with the other subject and Composite scores), rather than the previous scale from 2 to 12. The test also began asking test takers to give an opinion on a subject in light of three different perspectives provided by the test prompt, Lastly, the ACT began reporting four new "subscores," providing different ways to combine and evaluate the results of the various sections.

Beginning in September 2016, the scoring of the writing section changed back to a 2 to 12 scale.

ACT Mathematics Test

Number and Quantity

Numbers usually serve as an adjective representing a quantity of objects. They function as placeholders for a value. Numbers can be better understood by their type and related characteristics.

Integers

An **integer** is any number that does not have a fractional part. This includes all positive and negative **whole numbers** and zero. Fractions and decimals—which aren't whole numbers—aren't integers.

Prime Numbers

A **prime** number cannot be divided except by 1 and itself. A prime number has no other factors, which means that no other combination of whole numbers can be multiplied to reach that number. For example, the set of prime numbers between 1 and 27 is {2, 3, 5, 7, 11, 13, 17, 19, 23}.

The number 7 is a prime number because its only factors are 1 and 7. In contrast, 12 isn't a prime number, as it can be divided by other numbers like 2, 3, 4, and 6. Because they are composed of multiple factors, numbers like 12 are called **composite** numbers. All numbers greater than 1 that aren't prime numbers are composite numbers.

Even and Odd Numbers

An integer is **even** if one of its factors is 2, while those integers without a factor of 2 are **odd**. No numbers except for integers can have either of these labels. For example, 2, 40, -16, and 108 are all even numbers, while -1, 13, 59, and 77 are all odd numbers since they are integers that cannot be divided by 2 without a remainder. Numbers like 0.4, $\frac{5}{9}$, π, and $\sqrt{7}$ are neither odd nor even because they are not integers.

Decimals

A **decimal** number is designated by a **decimal point**, which indicates that what follows the point is a value that is less than 1 and is added to the integer number preceding the decimal point. The digit immediately following the decimal point is in the tenths place, the digit following the tenths place is in the hundredths place, and so on.

For example, the decimal number 1.735 has a value greater than 1 but less than 2. The 7 represents seven tenths of the unit 1 (0.7 or $\frac{7}{10}$); the 3 represents three hundredths of 1 (0.03 or $\frac{3}{100}$); and the 5 represents five thousandths of 1 (0.005 or $\frac{5}{1000}$).

Rational and Irrational Numbers

Rational numbers include all numbers that can be expressed as a fraction; in other words, rational numbers encompass all integers and all numbers with terminating or repeating decimals. That is, any rational number either will have a countable number of nonzero digits or will end with an ellipses or a bar (3.6666... or $3.\bar{6}$) to depict repeating decimal digits.

Some examples of rational numbers include 12, -3.54, $110.\overline{256}$, $\frac{-35}{10}$, and $4.\overline{7}$.

Irrational numbers include all real numbers that aren't rational. It can be thought of as any number with endless non-repeating digits to the right of the decimal point. They can be expressed as an endless decimal but never as a fraction. The most common irrational number is π, which has an endless and non-repeating decimal, but there are other well-known irrational numbers like e and $\sqrt{2}$.

Real Numbers

Defined by Descartes in the seventeenth century, **real numbers** include all numbers found on an infinite number line. All irrational and rational numbers are real numbers. Nonterminating decimal numbers and π are also real numbers. As the range of real numbers extends to both negative and positive infinity, the set of real numbers is complete and uncountable. This set is known as the **complete ordered field of numbers.**

Rounding Numbers

It's often convenient to **round** a number, which means to give an approximate figure to make it easier to compare amounts or perform mental math. When rounding to a certain place value, consider the next digit after that place value. When that digit is 5 or more, the digit in the selected place value gets rounded up. The digit used to determine the rounding, and all subsequent digits, become 0, and the selected place value is increased by 1. Here are some examples:

75 rounded to the nearest ten is 80

380 rounded to the nearest hundred is 400

22.697 rounded to the nearest hundredth is 22.70

When rounding to a certain place value, again consider the next digit after that place value. When that digit is below 5, the digit in the selected place value stays the same. The digit used to determine the rounding, and all subsequent digits, become 0.

Here are some examples:

92 rounded to the nearest ten is 90

839 rounded to the nearest hundred is 800

22.64 rounded to the nearest hundredth is 22.60

Addition

Addition is the combination of two numbers so their quantities are added together cumulatively. The sign for an addition operation is the + symbol. For example, 9 + 6 = 15. The 9 and 6 combine to achieve a cumulative value, called a **sum**.

Addition holds the **commutative property**, which means that numbers in an addition equation can be switched without altering the result. The formula for the commutative property is a + b = b + a. The following examples can demonstrate how the commutative property works:

$$7 = 3 + 4 = 4 + 3 = 7$$

$$20 = 12 + 8 = 8 + 12 = 20$$

Addition also holds the **associative property**, which means that the grouping of numbers does not matter in an addition problem. In other words, the presence or absence of parentheses is irrelevant. The formula for the associative property is (a + b) + c = a + (b + c). Here are some examples of the associative property at work:

$$30 = (6 + 14) + 10 = 6 + (14 + 10) = 30$$

$$35 = 8 + (2 + 25) = (8 + 2) + 25 = 35$$

Subtraction

Subtraction is taking away one number from another, so their quantities are reduced. The sign designating a subtraction operation is the − symbol, and the result is called the **difference.** For example, 9 - 6 = 3. The number *6* detracts from the number *9* to reach the difference *3*.

Unlike addition, subtraction follows neither the commutative nor associative properties. The order and grouping in subtraction impact the result.

$$15 = 22 - 7 \neq 7 - 22 = -15$$

$$3 = (10 - 5) - 2 \neq 10 - (5 - 2) = 7$$

When working through subtraction problems involving larger numbers, it's necessary to regroup the numbers. The following practice problem uses regrouping:

$$
\begin{array}{r}
3\;2\;5 \\
-\;7\;7 \\
\hline
\end{array}
$$

Here, it is clear that the ones and tens columns for 77 are greater than the ones and tens columns for 325. To subtract this number, one needs to borrow from the tens and hundreds columns. When borrowing from a column, subtracting 1 from the lender column will add 10 to the borrower column:

$$
\begin{array}{ccc}
3\text{-}1 & 10\text{+}2\text{-}1 & 10\text{+}5 \\
- & 7 & 7
\end{array}
=
\begin{array}{ccc}
2 & 11 & 15 \\
- & 7 & 7 \\
\hline
2 & 4 & 8
\end{array}
$$

After ensuring that each digit in the top row is greater than the digit in the corresponding bottom row, subtraction can proceed as normal, and the answer is found to be 248.

Multiplication

Multiplication involves adding together multiple copies of a number. It is indicated by an $\times$ symbol or a number immediately outside of a parenthesis. For example:

$$5(8 - 2)$$

The two numbers being multiplied together are called **factors**, and their result is called a **product**. For example, $9 \times 6 = 54$. This can be shown alternatively by expansion of either the 9 or the 6:

$$9 \times 6 = 9 + 9 + 9 + 9 + 9 + 9 = 54$$

$$9 \times 6 = 6 + 6 + 6 + 6 + 6 + 6 + 6 + 6 + 6 = 54$$

Like addition, multiplication holds the commutative and associative properties:

$$115 = 23 \times 5 = 5 \times 23 = 115$$

$$84 = 3 \times (7 \times 4) = (3 \times 7) \times 4 = 84$$

Multiplication also follows the **distributive property**, which allows the multiplication to be distributed through parentheses. The formula for distribution is $a \times (b + c) = ab + ac$. This is clear after the examples:

$$45 = 5 \times 9 = 5(3 + 6) = (5 \times 3) + (5 \times 6) = 15 + 30 = 45$$

$$20 = 4 \times 5 = 4(10 - 5) = (4 \times 10) - (4 \times 5) = 40 - 20 = 20$$

Multiplication becomes slightly more complicated when multiplying numbers with decimals. The easiest way to answer these problems is to ignore the decimals and multiply as if they were whole numbers. After multiplying the factors, a decimal gets placed in the product. The placement of the decimal is determined by taking the cumulative number of decimal places in the factors.

For example:

$$
\begin{array}{r}
0.7 \\
\times\ 3 \\
\hline
2.1
\end{array}
\qquad
\begin{array}{r}
2.6 \\
\times\ 4.2 \\
\hline
10.92
\end{array}
\qquad
\begin{array}{r}
1.5 \\
\times\ 6.4 \\
\hline
9.60
\end{array}
$$

Starting with the first example, the first step is to ignore the decimal and multiply the numbers as though they were whole numbers, which results in a product of 21. The next step is to count the number of digits that follow a decimal (one, in this case). Finally, the decimal place gets moved that many positions to the left, because the factors have only one decimal place. The second example works the same way, except that there are two total decimal places in the factors, so the product's decimal is moved two places over. In the third example, the decimal should be moved over two digits, but the digit zero is no longer needed, so it is erased and the final answer is 9.6.

Division

Division and multiplication are inverses of each other in the same way that addition and subtraction are opposites. The signs designating the division operation are the ÷ and / symbols. In division, the second number divides into the first.

The number before the division sign is called the **dividend** or, if expressed as a fraction, the **numerator.** For example, in $a \div b$, a is the dividend, while in $\frac{a}{b}$, a is the numerator.

The number after the division sign is called the **divisor** or, if expressed as a fraction, the **denominator.** For example, in $a \div b$, b is the divisor, while in $\frac{a}{b}$, b is the denominator.

Like subtraction, division doesn't follow the commutative property, as it matters which number comes before the division sign, and division doesn't follow the associative or distributive properties for the same reason. For example:

$$\frac{3}{2} = 9 \div 6 \neq 6 \div 9 = \frac{2}{3}$$

$$2 = 10 \div 5 = (30 \div 3) \div 5 \neq 30 \div (3 \div 5) = 30 \div \frac{3}{5} = 50$$

$$25 = 20 + 5 = (40 \div 2) + (40 \div 8) \neq 40 \div (2 + 8) = 40 \div 10 = 4$$

If a divisor doesn't divide into a dividend an integer number of times, whatever is left over is termed the **remainder**. The remainder can be further divided out into decimal form by using long division; however, this doesn't always give a **quotient** with a finite number of decimal places, so the remainder can also be expressed as a fraction over the original divisor.

Division with decimals is similar to multiplication with decimals in that when dividing a decimal by a whole number, one should ignore the decimal and divide as if it was a whole number.

Upon finding the answer, or quotient, the decimal point is inserted at the decimal place equal to that in the dividend.

$$15.75 \div 3 = 5.25$$

When the divisor is a decimal number, both the divisor and dividend get multiplied by 10. This process is repeated until the divisor is a whole number, then one needs to complete the division operation as described above.

$$17.5 \div 2.5 = 175 \div 25 = 7$$

Exponents

An **exponent** is an operation used as shorthand for a number multiplied or divided by itself for a defined number of times.

$$3^7 = 3 \times 3 \times 3 \times 3 \times 3 \times 3 \times 3$$

In this example, the 3 is called the **base** and the 7 is called the **exponent**. The exponent is typically expressed as a superscript number near the upper right side of the base but can also be identified as the number following a caret symbol (^). This operation is verbally expressed as "3 to the 7th power" or "3 raised to the power of 7." Common exponents are 2 and 3. A base raised to the power of 2 is referred to as having been "squared," while a base raised to the power of 3 is referred to as having been "cubed."

Several special rules apply to exponents. First, the **Zero Power Rule** finds that any number raised to the zero power equals 1. For example, 100^0, 2^0, $(-3)^0$ and 0^0 all equal 1 because the bases are raised to the zero power.

Second, exponents can be negative. With negative exponents, the equation is expressed as a fraction, as in the following example:

$$3^{-7} = \frac{1}{3^7} = \frac{1}{3 \times 3 \times 3 \times 3 \times 3 \times 3 \times 3}$$

Third, the **Power Rule** concerns exponents being raised by another exponent. When this occurs, the exponents are multiplied by each other:

$$(x^2)^3 = x^6 = (x^3)^2$$

Fourth, when multiplying two exponents with the same base, the **Product Rule** requires that the base remains the same and the exponents are added. For example, $a^x \times a^y = a^{x+y}$. Since addition and multiplication are commutative, the two terms being multiplied can be in any order.

$$x^3 x^5 = x^{3+5} = x^8 = x^{5+3} = x^5 x^3$$

Fifth, when dividing two exponents with the same base, the **Quotient Rule** requires that the base remains the same, but the exponents are subtracted. So, $a^x \div a^y = a^{x-y}$. Since subtraction and division are not commutative, the two terms must remain in order.

$$x^5 x^{-3} = x^{5-3} = x^2 = x^5 \div x^3 = \frac{x^5}{x^3}$$

Additionally, 1 raised to any power is still equal to 1, and any number raised to the power of 1 is equal to itself. In other words, $a^1 = a$ and $14^1 = 14$.

Exponents play an important role in scientific notation to present extremely large or small numbers as follows: $a \times 10^b$. To write the number in scientific notation, the decimal is moved until there is only one digit on the left side of the decimal point, indicating that the number a has a value between 1 and 10. The number of times the decimal moves indicates the exponent to which 10 is raised, here represented by b. If the decimal moves to the left, then b is positive, but if the decimal moves to the right, then b is negative. The following examples demonstrate these concepts:

$$3,050 = 3.05 \times 10^3$$

$$-777 = -7.77 \times 10^2$$

$$0.000123 = 1.23 \times 10^{-4}$$

$$-0.0525 = -5.25 \times 10^{-2}$$

Roots

The **square root** symbol is expressed as $\sqrt{}$ and is commonly known as the radical. Taking the root of a number is the inverse operation of multiplying that number by itself some number of times. For example, squaring the number 7 is equal to 7 × 7, or 49. Finding the square root is the opposite of finding an exponent, as the operation seeks a number that when multiplied by itself, equals the number in the square root symbol.

For example, $\sqrt{36} = 6$ because 6 multiplied by 6 equals 36. Note, the square root of 36 is also -6 since -6 × -6 = 36. This can be indicated using a **plus/minus** symbol like this: ±6. However, square roots are often just expressed as a positive number for simplicity, with it being understood that the true value can be either positive or negative.

Perfect squares are numbers with whole number square roots. The list of perfect squares begins with 0, 1, 4, 9, 16, 25, 36, 49, 64, 81, and 100.

Determining the square root of imperfect squares requires a calculator to reach an exact figure. It's possible, however, to approximate the answer by finding the two perfect squares that the number fits between. For example, the square root of 40 is between 6 and 7 since the squares of those numbers are 36 and 49, respectively.

Square roots are the most common root operation. If the radical doesn't have a number to the upper left of the symbol $\sqrt{}$, then it's a **square root**. Sometimes a radical includes a number in the upper left, like $\sqrt[3]{27}$, as in the other common root type—the **cube root**. Complicated roots, like the cube root, often require a calculator.

Parentheses

Parentheses separate different parts of an equation, and operations within them should be thought of as taking place before the outside operations take place. Practically, this means that the distinction between what is inside and outside of the parentheses decides the order of operations that the equation follows. Failing to solve operations inside the parentheses before addressing the part of the equation outside of the parentheses will lead to incorrect results.

For example, in $5 - (3 + 25)$, addition within the parentheses must be solved first. So $3 + 25 = 28$, leaving $5 - (28) = -23$. If this was solved using the incorrect order of operations, the solution might be found to be $5 - 3 + 25 = 2 + 25 = 27$, which would be wrong.

Equations often feature multiple layers of parentheses. To differentiate them, **square brackets** [ ] and **braces** { } are used in addition to parentheses. The innermost parentheses must be solved before working outward to larger brackets. For example, in $\{2 \div [5 - (3 + 1)]\}$, solving the innermost parentheses (3 + 1) leaves $\{2 \div [5 - (4)]\}$. $[5 - (4)]$ is now the next smallest, which leaves $\{2 \div [1]\}$ in the final step, and 2 as the answer.

Order of Operations

When solving equations with multiple operations, special rules apply. These rules are known as the **Order of Operations**. The order is as follows: Parentheses, Exponents, Multiplication and Division from left to right, and Addition and Subtraction from left to right. A popular mnemonic device to help remember the order is Please Excuse My Dear Aunt Sally (PEMDAS). Evaluating the following two problems can help with understanding the Order of Operations:

$$4 + (3 \times 2)^2 \div 4$$

First, the operation within the parentheses must be completed, yielding: $4 + 6^2 \div 4$.

Second, the exponent is evaluated: $4 + 36 \div 4$.

Third, the division is conducted: $4 + 9$.

Fourth, addition is completed, giving the answer: 13.

$$2 \times (6 + 3) \div (2 + 1)^2$$

$$2 \times 9 \div (3)^2$$

$$2 \times 9 \div 9$$

$$18 \div 9$$

$$2$$

Positive and Negative Numbers

<u>Signs</u>
Aside from 0, numbers can be either positive or negative. The sign for a positive number is the plus sign or the + symbol, while the sign for a negative number is minus sign or the − symbol. If a number has no designation, then it's assumed to be positive.

<u>Absolute Values</u>
Both positive and negative numbers are valued according to their distance from 0. Both +3 and -3 can be considered using the following number line:

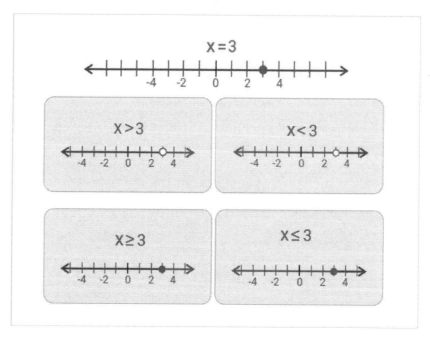

Both 3 and -3 are three spaces from 0. The distance from 0 is called its **absolute value**. Thus, both -3 and 3 have an absolute value of 3 since they're both three spaces away from 0.

An absolute number is written by placing | | around the number. So, |3| and |−3| both equal 3, as that's their common absolute value.

Implications for Addition and Subtraction

For addition, if all numbers are either positive or negative, they are simply added together. For example, 4 + 4 = 8 and -4 + -4 = -8. However, things get tricky when some of the numbers are negative and some are positive.

For example, with 6 + (-4), the first step is to take the absolute values of the numbers, which are 6 and 4. Second, the smaller value is subtracted from the larger. The equation becomes $6 - 4 = 2$. Third, the sign of the original larger number is placed on the sum. Here, 6 is the larger number, and it's positive, so the sum is 2.

Here's an example where the negative number has a larger absolute value: (-6) + 4. The first two steps are the same as the example above. However, on the third step, the negative sign must be placed on the sum, because the absolute value of (-6) is greater than 4. Thus, -6 + 4 = -2.

The absolute value of numbers implies that subtraction can be thought of as flipping the sign of the number following the subtraction sign and simply adding the two numbers. This means that subtracting a negative number will, in fact, be adding the positive absolute value of the negative number. Here are some examples:

$$-6 - 4 = -6 + -4 = -10$$

$$3 - -6 = 3 + 9 = 12$$

$$-3 - 2 = -3 + -2 = -5$$

Implications for Multiplication and Division

For multiplication and division, if both numbers are positive, then the product or quotient is always positive. If both numbers are negative, then the product or quotient is also positive. However, if the numbers have opposite signs, the product or quotient is always negative.

Simply put, the product in multiplication and quotient in division is always positive, unless the numbers have opposing signs, in which case it's negative. Here are some examples:

$$(-6) \times (-5) = 30$$

$$(-50) \div 10 = -5$$

$$8 \times |-7| = 56$$

$$(-48) \div (-6) = 8$$

If there are more than two numbers in a multiplication or division problem, then whether the product or quotient is positive or negative depends on the number of negative numbers in the problem. If there is an odd number of negatives, then the product or quotient is negative. If there is an even number of negative numbers, then the result is positive.

Here are some examples:

$$(-6) \times 5 \times (-2) \times (-4) = -240$$

$$(-6) \times 5 \times 2 \times (-4) = 240$$

Factorization

Factors are the numbers multiplied to achieve a product. Thus, every product in a multiplication equation has, at minimum, two factors. Of course, some products will have more than two factors. For the sake of most discussions, one can assume that factors are positive integers.

To find a number's factors, one should start with 1 and the number itself. The next step is to divide the number by 2, 3, 4, and so on, to see if any divisors can divide the number without a remainder. A list should be kept of those that do. This process can be stopped upon reaching either the number itself or another factor.

For example, to find the factors of 45, the first step is to start with 1 and 45. The next step is to try to divide 45 by 2, which fails. After this, 45 gets divided by 3. The answer is 15, so 3 and 15 are now factors. Dividing by 4 doesn't work and dividing by 5 leaves 9. Lastly, dividing 45 by 6, 7, and 8 all don't work. The next integer to try is 9, but this is already known to be a factor, so the factorization is complete. The factors of 45 are 1, 3, 5, 9, 15 and 45.

Prime Factorization

Prime factorization involves an additional step after breaking a number down to its factors: breaking down the factors until they are all prime numbers. A **prime number** is any number that can only be divided by 1 and itself. The prime numbers between 1 and 20 are 2, 3, 5, 7, 11, 13, 17, and 19. As a simple test, numbers that are even or end in 5 are not prime.

When attempting to break 129 down into its prime factors, the factors are found first: 3 and 43. Both 3 and 43 are prime numbers, so that means the prime factorization is complete. If 43 was not a prime number, then it would also need to be factorized until all of the factors were expressed as prime numbers.

Common Factor

A **common factor** is a factor shared by two numbers. The following examples demonstrate how to find the common factors of 45 and 30:

- o The factors of 45 are: 1, 3, 5, 9, 15, and 45.
- o The factors of 30 are: 1, 2, 3, 5, 6, 10, 15, and 30.
- o The common factors are 1, 3, 5, and 15.

Greatest Common Factor

The **greatest common factor** is the largest number among the shared, common factors. From the factors of 45 and 30, the common factors are 3, 5, and 15. Thus, 15 is the greatest common factor, as it's the largest number.

Least Common Multiple

The **least common multiple** is the smallest number that's a multiple of two numbers. For example, to find the least common multiple of 4 and 9, the multiples of 4 and 9 are found first. The multiples of 4 are 4, 8, 12, 16, 20, 24, 28, 32, 36, and so on. For 9, the multiples are 9, 18, 27, 36, 45, 54, etc. Thus, the least common multiple of 4 and 9 is 36, the lowest number where 4 and 9 share multiples.

If two numbers share no factors besides 1 in common, then their least common multiple will be simply their product. If two numbers have common factors, then their least common multiple will be their product divided by their greatest common factor. This can be visualized by the formula $LCM = \frac{x \times y}{GCF}$, where x and y are some integers and LCM and GCF are their least common multiple and greatest common factor, respectively.

Fractions

A **fraction** is an equation that represents a part of a whole but can also be used to present ratios or division problems. An example of a fraction is $\frac{x}{y}$. In this example, x is called the **numerator,** while y is the **denominator.** The numerator represents the number of parts, and the denominator is the total number of parts. They are separated by a line or slash, known as a fraction bar.. In simple fractions, the numerator and denominator can be nearly any integer. However, the denominator of a fraction can never be zero, because dividing by zero is a function, which is undefined.

To visualize the basic idea of fractions, one can imagine that an apple pie has been baked for a holiday party, and the full pie has eight slices. After the party, there are five slices left. How could the amount of the pie that remains be expressed as a fraction? The numerator is 5 since there are five parts left, and the denominator is 8, since there were eight total slices in the whole pie. Thus, expressed as a fraction, the leftover pie totals $\frac{5}{8}$ of the original amount.

Fractions come in three different varieties: proper fractions, improper fractions, and mixed numbers. **Proper fractions** have a numerator less than the denominator, such as $\frac{3}{8}$, but **improper fractions** have a numerator greater than the denominator, such as $\frac{15}{8}$. **Mixed numbers** combine a whole number with a proper fraction, such as $3\frac{1}{2}$. Any mixed number can be written as an improper fraction by multiplying the integer by the denominator, adding the product to the value of the numerator, and dividing the sum by the original denominator. For example, $3\frac{1}{2} = \frac{3 \times 2 + 1}{2} = \frac{7}{2}$. Whole numbers can also be converted into fractions by placing the whole number as the numerator and making the denominator 1. For example, $3 = \frac{3}{1}$.

One of the most fundamental concepts of fractions is their ability to be manipulated by multiplication or division. This is possible since $\frac{n}{n} = 1$ for any non-zero integer. As a result, multiplying or dividing by $\frac{n}{n}$ will not alter the original fraction since any number multiplied or divided by 1 doesn't change the value of that number. Fractions of the same value are known as equivalent fractions. For example, $\frac{2}{4}, \frac{4}{8}, \frac{50}{100}$, and $\frac{75}{150}$ are equivalent, as they all equal $\frac{1}{2}$.

Although many equivalent fractions exist, they are easier to compare and interpret when reduced or simplified. The numerator and denominator of a simple fraction will have no factors in common other than 1. When reducing or simplifying fractions, the numerator and denominator are divided by the greatest common factor. A simple strategy is to divide the numerator and denominator by low numbers, like 2, 3, or 5 until arriving at a simple fraction, but the same thing could be achieved by determining the greatest common factor for both the numerator and denominator and dividing each by it. Using the first method is preferable when both the numerator and denominator are even, end in 5, or are obviously a multiple of another number. However, if no numbers seem to work, it will be necessary to factor the numerator and denominator to find the GCF. The following problems provide examples:

Simplify the fraction $\frac{6}{8}$:

Dividing the numerator and denominator by 2 results in $\frac{3}{4}$, which is a simple fraction.

Simplify the fraction $\frac{12}{36}$:

Dividing the numerator and denominator by 2 leaves $\frac{6}{18}$. This isn't a simplified fraction, as both the numerator and denominator have factors in common. Diving each by 3 results in $\frac{2}{6}$, but this can be further simplified by dividing by 2, to get $\frac{1}{3}$. This is the simplest fraction, as the numerator is 1. In cases like this, multiple division operations can be avoided by determining the greatest common factor between the numerator and denominator.

Simplify the fraction $\frac{18}{54}$ by dividing by the greatest common factor:

The first step is to determine the factors of the numerator and denominator. The factors of 18 are 1, 2, 3, 6, 9, and 18. The factors of 54 are 1, 2, 3, 6, 9, 18, 27, and 54. Thus, the greatest common factor is 18. Dividing $\frac{18}{54}$ by 18 leaves $\frac{1}{3}$, which is the simplest fraction. This method takes slightly more work, but it definitively arrives at the simplest fraction.

A **ratio** is a comparison between the relative sizes of two parts of a whole, separated by a colon. It's different from a fraction because, in a ratio, the second number represents the number of parts which aren't currently being referenced, while in a fraction, the second or bottom number represents the total number of parts in the whole. For example, if 3 pieces of an 8-piece pie were eaten, the number of uneaten parts expressed as a ratio to the number of eaten parts would be 5:3.

Equivalent ratios work just like equivalent fractions. For example, both 3:9 and 20:60 are equivalent ratios to 1:3 because both can be simplified to 1:3.

Operations with Fractions

Of the four basic operations that can be performed on fractions, the one that involves the least amount of work is multiplication. To multiply two fractions, the numerators are multiplied, the denominators are multiplied, and the products are placed together as a fraction. Whole numbers and mixed numbers can also be expressed as a fraction, as described above, which more easily facilitates multiplication with another fraction. The following problems provide examples:

1. $\frac{2}{5} \times \frac{3}{4} = \frac{6}{20} = \frac{3}{10}$

2. $\frac{4}{9} \times \frac{7}{11} = \frac{28}{99}$

Dividing fractions is similar to multiplication with one key difference. To divide fractions, the numerator and denominator of the second fraction are flipped, and then one proceeds as if it were a multiplication problem:

1. $\frac{7}{8} \div \frac{4}{5} = \frac{7}{8} \times \frac{5}{4} = \frac{35}{32}$

2. $\frac{5}{9} \div \frac{1}{3} = \frac{5}{9} \times \frac{3}{1} = \frac{15}{9} = \frac{5}{3}$

Addition and subtraction require more steps than multiplication and division, as these operations require the fractions to have the same denominator, also called a **common denominator**. It is always possible to find a common denominator by multiplying the denominators. However, when the denominators are large numbers, this method is unwieldy, especially if the answer must be provided in its simplest form. Thus, it's beneficial to find the least common denominator of the fractions—the least common denominator is incidentally also the least common multiple.

Once equivalent fractions have been found with common denominators, the numerators are simply added or subtracted to arrive at the answer:

1. $\frac{1}{2} + \frac{3}{4} = \frac{2}{4} + \frac{3}{4} = \frac{5}{4}$

2. $\frac{3}{12} + \frac{11}{20} = \frac{15}{60} + \frac{33}{60} = \frac{48}{60} = \frac{4}{5}$

3. $\frac{7}{9} - \frac{4}{15} = \frac{35}{45} - \frac{12}{45} = \frac{23}{45}$

4. $\frac{5}{6} - \frac{7}{18} = \frac{15}{18} - \frac{7}{18} = \frac{8}{18} = \frac{4}{9}$

Order of Rational Numbers

A common question type on the ACT asks test takers to order rational numbers from least to greatest or greatest to least. The numbers will come in a variety of formats, including decimals, percentages, roots, fractions, and whole numbers. These questions test for knowledge of different types of numbers and the ability to determine their respective values.

Whether the question asks to order the numbers from greatest to least or least to greatest, the crux of the question is the same—convert the numbers into a common format. Generally, it's easiest to write the numbers as whole numbers and decimals so they can be placed on a number line. The following examples illustrate this strategy:

1. Order the following rational numbers from greatest to least:

$$\sqrt{36}, 0.65, 78\%, \frac{3}{4}, 7, 90\%, \frac{5}{2}$$

Of the seven numbers, the whole number (7) and decimal (0.65) are already in an accessible form, so test takers should concentrate on the other five.

First, the square root of 36 equals 6. (If the test asks for the root of a non-perfect root, determine which two whole numbers the root lies between.) Next, the percentages should be converted to decimals. A percentage means "per hundred," so this conversion requires moving the decimal point two places to the left, leaving 0.78 and 0.9. Lastly, the fractions are evaluated: $\frac{3}{4} = \frac{75}{100} = 0.75$; $\frac{5}{2} = 2\frac{1}{2} = 2.5$

Now, the only step left is to list the numbers in the requested order:

$$7, \sqrt{36}, \frac{5}{2}, 90\%, 78\%, \frac{3}{4}, 0.65$$

2. Order the following rational numbers from least to greatest:

$$2.5, \sqrt{9}, -10.5, 0.853, 175\%, \sqrt{4}, \frac{4}{5}$$

$$\sqrt{9} = 3$$

$$175\% = 1.75$$

$$\sqrt{4} = 2$$

$$\frac{4}{5} = 0.8$$

From least to greatest, the answer is: $-10.5, \frac{4}{5}, 0.853, 175\%, \sqrt{4}, 2.5, \sqrt{9},$

Percentages

Percentages can be thought of as fractions with a denominator of 100. In fact, percentage means "per hundred." Problems often require converting numbers from percentages, fractions, and decimals. The following explains how to work through those conversions.

Converting Fractions to Percentages: The fraction is converted by using an equivalent fraction with a denominator of 100. For example, $\frac{3}{4} = \frac{3}{4} \times \frac{25}{25} = \frac{75}{100} = 75\%$

Converting Percentages to Fractions: Percentages can be converted to fractions by turning the percentage into a fraction with a denominator of 100. Test takers should be wary of questions asking the converted fraction to be written in the simplest form. For example, $35\% = \frac{35}{100}$ which, although correctly written, has a numerator and denominator with a greatest common factor of 5, so it can be simplified to $\frac{7}{20}$.

Converting Percentages to Decimals: Because a percentage is based on "per hundred," decimals and percentages can be converted by multiplying or dividing by 100. Practically speaking, this always amounts to moving the decimal point two places to the right or left, depending on the conversion. To convert a percentage to a decimal, the decimal point is moved two places to the left and the % sign gets removed. To convert a decimal to a percentage, the decimal point is moved two places to the right and a "%" sign is added. Here are some examples:

65% = 0.65

0.33 = 33%

0.215 = 21.5%

99.99% = 0.9999

500% = 5.00

7.55 = 755%

Percentage Problems

Questions dealing with percentages can be difficult when they are phrased as word problems. These word problems almost always come in one of three varieties. The first type will ask to find what percentage of some number will equal another number. The second asks to determine what number is some percentage of another given number. The third will ask what number another number is a given percentage of.

One of the most important parts of correctly answering percentage word problems is to identify the numerator and the denominator. This fraction can then be converted into a percentage, as described in the previous section.

The following word problem shows how to make this conversion:

A department store carries several different types of footwear. The store is currently selling 8 athletic shoes, 7 dress shoes, and 5 sandals. What percentage of the store's footwear are sandals?

The first step is to calculate what serves as the 'whole', as this will be the denominator. How many total pieces of footwear does the store sell? The store sells 20 different types (8 athletic + 7 dress + 5 sandals).

In the next step, test takers need to determine which footwear type the question is specifically asking about. Sandals. Thus, 5 is the numerator.

Lastly, the resultant fraction must be expressed as a percentage. The first two steps indicate that $\frac{5}{20}$ of the footwear pieces are sandals. This fraction must now be converted into a percentage: $\frac{5}{20} \times \frac{5}{5} = \frac{25}{100} = 25\%$.

Vectors

A **vector** can be thought of as an abstract list of numbers or as giving a location in a space. For example, the coordinates (x, y) for points in the Cartesian plane are vectors. Each entry in a vector can be referred to by its location in the list: first, second, third, and so on. The total length of the list is the **dimension** of the vector. A vector is often denoted as such by putting an arrow on top of it, e.g. $\vec{v} = (v_1, v_2, v_3)$.

Adding Vectors Graphically and Algebraically

There are two basic operations for vectors. First, two vectors can be added together. Let $\vec{v} = (v_1, v_2, v_3), \vec{w} = (w_1, w_2, w_3)$. The the sum of the two vectors is defined to be $\vec{v} + \vec{w} = (v_1 + w_1, v_2 + w_2, v_3 + w_3)$. Subtraction of vectors can be defined similarly.

Vector addition can be visualized in the following manner. First, each vector can be visualized as an arrow. Then, the base of one arrow is placed at the tip of the other arrow. The tip of this first arrow now hits some point in space, and there will be an arrow from the origin to this point. This new arrow corresponds to the new vector. In subtraction, the direction of the arrow being subtracted is reversed.

For example, if adding together the vectors (-2, 3) and (4, 1), the new vector will be (-2 + 4, 3 + 1), or (2, 4). Graphically, this may be pictured in the following manner:

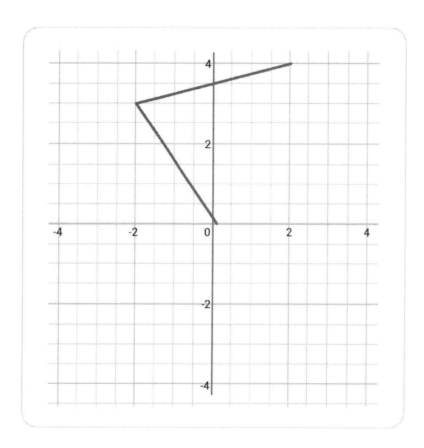

Performing Scalar Multiplications

The second basic operation for vectors is called **scalar multiplication**. Scalar multiplication is multiplying any vector by any real number, which is denoted here as a scalar. Let $\vec{v} = (v_1, v_2, v_3)$, and let a be an arbitrary real number. Then the scalar multiple $a\vec{v} = (av_1, av_2, av_3)$. Graphically, this corresponds to changing the length of the arrow corresponding to the vector by a factor, or scale, of a. That is why the real number is called a **scalar** in this instance.

As an example, let $\vec{v} = (2, -1, 1)$. Then $3\vec{v} = (3 \cdot 2, 3(-1), 3 \cdot 1) = (6, -3, 3)$.

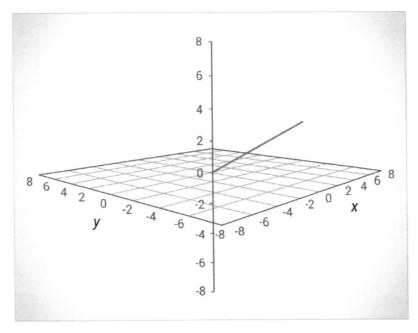

Note that scalar multiplication is **distributive** over vector addition, meaning that $a(\vec{v} + \vec{w}) = a\vec{v} + a\vec{w}$.

Determinants

A **matrix** is a rectangular arrangement of numbers in rows and columns. The **determinant** of a matrix is a special value that can be calculated for any square matrix.

Using the *square 2 x 2 matrix* $\begin{bmatrix} a & b \\ c & d \end{bmatrix}$, the determinant is $ad - bc$.

For example, the determinant of the matrix $\begin{bmatrix} -5 & 1 \\ 3 & 4 \end{bmatrix}$ is *-5(4) – 1(3) = -20 – 3 = -23*.

Using a *3 x 3 matrix* $\begin{bmatrix} a & b & c \\ d & e & f \\ g & h & i \end{bmatrix}$, the determinant is $a(ei - fh) - b(di - fg) + c(dh - eg)$.

For example, the determinant of the matrix $\begin{bmatrix} 2 & 0 & 1 \\ -1 & 3 & 2 \\ 2 & -2 & -1 \end{bmatrix}$ is

$$2\big(3(-1) - 2(-2)\big) - 0\big(-1(-1) - 2(2)\big) + 1\big(-1(-2) - 3(2)\big)$$

$$= 2(-3 + 4) - 0(1 - 4) + 1(2 - 6)$$

$$= 2(1) - 0(-3) + 1(-4)$$

$$= 2 - 0 - 4 = -2$$

The pattern continues for larger square matrices.

25

Algebra

Solving for X in Proportions

Proportions are commonly used in word problems to find unknown values, such as x, that are some percent or fraction of a known number. Proportions are solved by cross-multiplying and then dividing to arrive at x. The following examples show how this is done:

$$1. \quad \frac{75\%}{90\%} = \frac{25\%}{x}$$

To solve for x, the fractions must be cross-multiplied: $(75\%x = 90\% \times 25\%)$. To make things easier, the percentages can be converted to decimals: $(0.9 \times 0.25 = 0.225 = 0.75x)$. To get rid of the coefficient of x, each side must be divided by that same coefficient to get the answer $x = 0.3$. The question could ask for the answer as a percentage or fraction in lowest terms, which are 30% and $\frac{3}{10}$, respectively.

$$2. \quad \frac{x}{12} = \frac{30}{96}$$

Cross-multiply: $96x = 30 \times 12$

Multiply: $96x = 360$

Divide: $x = 360 \div 96$

Answer: $x = 3.75$

$$3. \quad \frac{0.5}{3} = \frac{x}{6}$$

Cross-multiply: $3x = 0.5 \times 6$

Multiply: $3x = 3$

Divide: $x = 3 \div 3$

Answer: $x = 1$

Observant test takers may have noticed there's a faster way to arrive at the answer. If there is an obvious operation being performed on the proportion, the same operation can be used on the other side of the proportion to solve for x. For example, in the first practice problem, 75% became 25% when divided by 3, and upon doing the same to 90%, the correct answer of 30% would have been found with much less legwork. However, these questions aren't always so intuitive, so it's a good idea for test takers to work through the steps, even if the answer seems apparent from the onset.

Translating Words into Math

If this was solved in the incorrect order of operations as a situation or translated from a word problem into an expression, test takers should look for key words indicating addition, subtraction, multiplication, or division:

- o *Addition*: add, altogether, together, plus, increased by, more than, in all, sum, and total
- o *Subtraction*: minus, less than, difference, decreased by, fewer than, remain, and take away
- o *Multiplication*: times, twice, of, double, and triple
- o *Division*: divided by, cut up, half, quotient of, split, and shared equally

If a question asks to give words to a mathematical expression and says "equals," then an = sign must be included in the answer. Similarly, "less than or equal to" is expressed by the inequality symbol $\leq$, and "greater than or equal" to is expressed as $\geq$. Furthermore, "less than" is represented by <, and "greater than" is expressed by >.

Word Problems

Word problems can appear daunting, but prepared test takers shouldn't let the verbiage psyche them out. No matter the scenario or specifics, the key to answering them is to translate the words into a math problem. It is critical to keep in mind what the question is asking and what operations could lead to that answer. The following word problems highlight the most commonly tested question types.

Working with Money
Walter's Coffee Shop sells a variety of drinks and breakfast treats.

Price List	
Hot Coffee	$2.00
Slow Drip Iced Coffee	$3.00
Latte	$4.00
Muffins	$2.00
Crepe	$4.00
Egg Sandwich	$5.00

Costs	
Hot Coffee	$0.25
Slow Drip Iced Coffee	$0.75
Latte	$1.00
Muffins	$1.00
Crepe	$2.00
Egg Sandwich	$3.00

Walter's utilities, rent, and labor costs him $500 per day. Today, Walter sold 200 hot coffees, 100 slow drip iced coffees, 50 lattes, 75 muffins, 45 crepes, and 60 egg sandwiches. What was Walter's total profit today?

To accurately answer this type of question, the first step is to determine the total cost of making his drinks and treats, then determine how much revenue he earned from selling those products. After arriving at these two totals, the profit is measured by deducting the total cost from the total revenue.

Walter's costs for today:

200 hot coffees	× $0.25	= $50
100 slow drip iced coffees	× $0.75	= $75
50 lattes	× $1.00	= $50
75 muffins	× $1.00	= $75
45 crepes	× $2.00	= $90
60 egg sandwiches	× $3.00	= $180
Utilities, Rent, and Labor		= $500
Total costs		= $1,020

Walter's revenue for today:

200 hot coffees	× $2.00	= $400
100 slow drip iced coffees	× $3.00	= $300
50 lattes	× $4.00	= $200
75 muffins	× $2.00	= $150
45 crepes	× $4.00	= $180
60 egg sandwiches	× $5.00	= $300
Total revenue		= $1,530

Walter's $Profit = Revenue - Costs = \$1,530 - \$1,020 = \$510$

This strategy can be applied to other question types. For example, calculating salary after deductions, balancing a checkbook, and calculating a dinner bill are common word problems similar to business planning. In all cases, the most important step is remembering to use the correct operations. When a balance is increased, addition is used. When a balance is decreased, the problem requires subtraction. Common sense and organization are one's greatest assets when answering word problems.

Unit Rate

Unit rate word problems ask test takers to calculate the rate or quantity of something in a different value. For example, a problem might say that a car drove a certain number of miles in a certain number of minutes and then ask how many miles per hour the car was traveling. These questions involve solving proportions. Consider the following examples:

1. Alexandra made $96 during the first 3 hours of her shift as a temporary worker at a law office. She will continue to earn money at this rate until she finishes in 5 more hours. How much does Alexandra make per hour? How much money will Alexandra have made at the end of the day?

This problem can be solved in two ways. The first is to set up a proportion, as the rate of pay is constant. The second is to determine her hourly rate, multiply the 5 hours by that rate, and then adding the $96.

To set up a proportion, the money already earned (numerator) is placed over the hours already worked (denominator) on one side of an equation. The other side has x over 8 hours (the total hours worked in the day). It looks like this: $\frac{96}{3} = \frac{x}{8}$. Now, cross-multiply yields $768 = 3x$. To get x, the 768 is divided by 3, which leaves $x = 256$. Alternatively, as x is the numerator of one of the proportions, multiplying by its denominator will reduce the solution by one step. Thus, Alexandra will make $256 at the end of the day. To calculate her hourly rate, the total is divided by 8, giving $32 per hour.

Alternatively, it is possible to figure out the hourly rate by dividing $96 by 3 hours to get $32 per hour. Now her total pay can be figured by multiplying $32 per hour by 8 hours, which comes out to $256.

2. Jonathan is reading a novel. So far, he has read 215 of the 335 total pages. It takes Jonathan 25 minutes to read 10 pages, and the rate is constant. How long does it take Jonathan to read one page? How much longer will it take him to finish the novel? Express the answer in time.

To calculate how long it takes Jonathan to read one page, 25 minutes is divided by 10 pages to determine the page per minute rate. Thus, it takes 2.5 minutes to read one page.

Jonathan must read 120 more pages to complete the novel. (This is calculated by subtracting the pages already read from the total.) Now, his rate per page is multiplied by the number of pages. Thus, $120 \times 2.5 = 300$. Expressed in time, 300 minutes is equal to 5 hours.

3. At a hotel, $\frac{4}{5}$ of the 120 rooms are booked for Saturday. On Sunday, $\frac{3}{4}$ of the rooms are booked. On which day are more of the rooms booked, and by how many more?

The first step is to calculate the number of rooms booked for each day. This is done by multiplying the fraction of the rooms booked by the total number of rooms.

$$\text{Saturday:} \frac{4}{5} \times 120 = \frac{4}{5} \times \frac{120}{1} = \frac{480}{5} = 96 \text{ rooms}$$

$$\text{Sunday:} \frac{3}{4} \times 120 = \frac{3}{4} \times \frac{120}{1} = \frac{360}{4} = 90 \text{ rooms}$$

Thus, more rooms were booked on Saturday by 6 rooms.

4. In a veterinary hospital, the veterinarian-to-pet ratio is 1:9. The ratio is always constant. If there are 45 pets in the hospital, how many veterinarians are currently in the veterinary hospital?

A proportion is set up to solve for the number of veterinarians: $\frac{1}{9} = \frac{x}{45}$

Cross-multiplying results in $9x = 45$, which works out to 5 veterinarians.

Alternatively, as there are always 9 times as many pets as veterinarians, is it possible to divide the number of pets (45) by 9. This also arrives at the correct answer of 5 veterinarians.

5. At a general practice law firm, 30% of the lawyers work solely on tort cases. If 9 lawyers work solely on tort cases, how many lawyers work at the firm?

The first step is to solve for the total number of lawyers working at the firm, which will be represented here with x. The problem states that 9 lawyers work solely on torts cases, and they make up 30% of the total lawyers at the firm. Thus, 30% multiplied by the total, x, will equal 9. Written as equation, this is: $30\% \times x = 9$.

It's easier to deal with the equation after converting the percentage to a decimal, leaving $0.3x = 9$. Thus, $x = \frac{9}{0.3} = 30$ lawyers working at the firm.

6. Xavier was hospitalized with pneumonia. He was originally given 35mg of antibiotics. Later, after his condition continued to worsen, Xavier's dosage was increased to 60mg. What was the percent increase of the antibiotics? Round the percentage to the nearest tenth.

An increase or decrease in percentage can be calculated by dividing the difference in amounts by the original amount and multiplying by 100. Written as an equation, the formula is:

$$\frac{new\ quantity\ -\ old\ quantity}{old\ quantity} \times 100$$

Here, the question states that the dosage was increased from 35mg to 60mg, so these values are plugged into the formula to find the percentage increase.

$$\frac{60\ -\ 35}{35} \times 100 = \frac{25}{35} \times 100 = .7142 \times 100 = 71.4\%$$

FOIL Method

FOIL is a technique for generating polynomials through the multiplication of binomials. A **polynomial** is an expression of multiple variables (for example, x, y, z) in at least three terms involving only the four basic operations and exponents. FOIL is an acronym for First, Outer, Inner, and Last. "First" represents the multiplication of the terms appearing first in the binomials. "Outer" means multiplying the outermost terms. "Inner" means multiplying the terms inside. "Last" means multiplying the last terms of each binomial.

After completing FOIL and solving the operations, **like terms** are combined. To identify like terms, test takers should look for terms with the same variable and the same exponent. For example, in $4x^2 - x^2 + 15x + 2x^2 - 8$, the $4x^2, -x^2$, and $2x^2$ are all like terms because they have the variable (x) and exponent (2). Thus, after combining the like terms, the polynomial has been simplified to $5x^2 + 15x - 8$.

The purpose of FOIL is to simplify an equation involving multiple variables and operations. Although it sounds complicated, working through some examples will provide some clarity:

    1. Simplify $(x + 10)(x + 4) =$

$(x \times x)$	$+$	$(x \times 4)$	$+$	$(10 \times x)$	$+$	(10×4)
First		Outer		Inner		Last

After multiplying these binomials, it's time to solve the operations and combine like terms. Thus, the expression becomes: $2x^2 + 4x + 10x + 40 = 2x^2 + 14x + 40$.

    2. Simplify $2x(4x^3 - 7y^2 + 3x^2 + 4)$

Here, a monomial $(2x)$ is multiplied into a polynomial $(4x^3 - 7y^2 + 3x^2 + 4)$. Using the distributive property, the monomial gets multiplied by each term in the polynomial. This becomes $2x(4x^3) - 2x(7y^2) + 2x(3x^2) + 2x(4)$.

Now, each monomial is simplified, starting with the coefficients:

$$(2 \times 4)(x \times x^3) - (2 \times 7)(x \times y^2) + (2 \times 3)(x \times x^2) + (2 \times 4)(x)$$

When multiplying powers with the same base, their exponents are added. Remember, a variable with no listed exponent has an exponent of 1, and exponents of distinct variables cannot be combined. This produces the answer:

$$8x^{1+3} - 14xy^2 + 6x^{1+2} + 8x = 8x^4 - 14xy^2 + 6x^3 + 8x$$

3. Simplify $(8x^{10}y^2z^4) \div (4x^2y^4z^7)$

The first step is to divide the coefficients of the first two polynomials: $8 \div 4 = 2$. The second step is to divide exponents with the same variable, which requires subtracting the exponents. This results in: $2(x^{10-2}y^{2-4}z^{4-7}) = 2x^8y^{-2}z^{-3}$.

However, the most simplified answer should include only positive exponents. Thus, $y^{-2}z^{-3}$ needs to be converted into fractions, respectively $\frac{1}{y^2}$ and $\frac{1}{z^3}$. Since the $2x^8$ has a positive exponent, it is placed in the numerator, and $\frac{1}{y^2}$ and $\frac{1}{z^3}$ are combined into the denominator, leaving $\frac{2x^8}{y^2z^3}$ as the final answer.

Rational Expressions

A **rational expression** is a fraction where the numerator and denominator are both polynomials. Some examples of rational expressions include the following: $\frac{4x^3y^5}{3z^4}$, $\frac{4x^3+3x}{x^2}$, and $\frac{x^2+7x+1}{x+2}$. Since these refer to expressions and not equations, they can be simplified but not solved. Using the rules in the previous *Exponents* and *Roots* sections, some rational expressions with monomials can be simplified. Other rational expressions such as the last example, $\frac{x^2+7x+10}{x+2}$, require more steps to be simplified. First, the polynomial on top can be factored from $x^2 + 7x + 10$ into $(x + 5)(x + 2)$. Then the common factors can be canceled and the expression can be simplified to $(x + 5)$.

The following problem is an example of using rational expressions:

Reggie wants to lay sod in his rectangular backyard. The length of the yard is given by the expression $4x + 2$ and the width is unknown. The area of the yard is $20x + 10$. Reggie needs to find the width of the yard. Knowing that the area of a rectangle is length multiplied by width, an expression can be written to find the width: $\frac{20x+1}{4x+2}$, area divided by length. Simplifying this expression by factoring out 10 on the top and 2 on the bottom leads to this expression: $\frac{10(2x+1)}{2(2x+1)}$. Cancelling out the $2x + 1$ results in $\frac{10}{2} = 5$. The width of the yard is found to be 5 by simplifying the rational expression.

Rational Equations

A **rational equation** can be as simple as an equation with a ratio of polynomials, $\frac{p(x)}{q(x)}$, set equal to a value, where $p(x)$ and $q(x)$ are both polynomials. A rational equation has an equal sign, which is different from expressions. This leads to solutions, or numbers that make the equation true.

It is possible to solve rational equations by trying to get all of the x terms out of the denominator and then isolating them on one side of the equation. For example, to solve the equation $\frac{3x+2}{2x+3} = 4$, both sides get multiplied by $(2x + 3)$. This will cancel on the left side to yield $3x + 2 = 4(2x + 3)$, then $3x + 2 = 8x + 12$. Now, subtract $8x$ from both sides, which yields $-5x + 2 = 12$. Subtracting 2 from both sides results in $-5x = 10$. Finally, both sides get divided by -5 to obtain $x = -2$.

Sometimes, when solving rational equations, it can be easier to try to simplify the rational expression by factoring the numerator and denominator first, then cancelling out common factors. For example, to solve $\frac{2x^2-8x+6}{x^2-3x+2} = 1$, the first step is to factor $2x^2 - 8x + 6 = 2(x^2 - 4x + 3) = 2(x - 1)(x - 3)$. Then, factor $x^2 - 3x + 2$ into $(x - 1)(x - 2)$. This turns the original equation into $\frac{2(x-1)(x-3)}{(x-1)(x-2)} = 1$. The common factor of $(x - 1)$ can be canceled, leaving $\frac{2(x-3)}{x-1} = 1$. Now the same method used in the previous example can be followed. Multiplying both sides by $x - 1$ and performing the multiplication on the left yields $2x - 6 = x - 1$, which can be simplified to $x = 5$.

Rational Functions

A **rational function** is similar to an equation, but it includes two variables. In general, a rational function is in the form: $f(x) = \frac{p(x)}{q(x)}$, where $p(x)$ and $q(x)$ are polynomials. Refer to the *Functions* section (which follows) for a more detailed definition of functions. Rational functions are defined everywhere except where the denominator is equal to zero. When the denominator is equal to zero, this indicates either a hole in the graph or an asymptote. An example of a function with an asymptote is shown below.

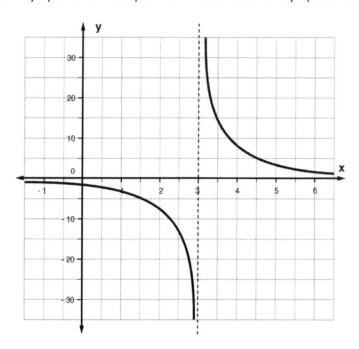

Functions

Algebraic Functions

A function is called **algebraic** if it is built up from polynomials by adding, subtracting, multiplying, dividing, and taking radicals. This means that, for example, the variable can never appear in an exponent. Thus, polynomials and rational functions are algebraic, but exponential functions are not algebraic. It turns out that logarithms and trigonometric functions are not algebraic either.

A function of the form $f(x) = a_n x^n + a_{n-1} x^{n-1} + a_{n-2} x^{n-2} + \cdots + a_1 x + a_0$ is called a **polynomial function**. The value of n is called the **degree** of the polynomial. In the case where $n = 1$, it is called a **linear function**. In the case where $n = 2$, it is called a **quadratic function**. In the case where $n = 3$, it is called a **cubic function**.

When n is even, the polynomial is called **even**, and not all real numbers will be in its range. When n is odd, the polynomial is called **odd**, and the range includes all real numbers.

The graph of a quadratic function $f(x) = ax^2 + bx + c$ will be a **parabola**. To see whether or not the parabola opens up or down, it's necessary to check the coefficient of x^2, which is the value a. If the coefficient is positive, then the parabola opens upward. If the coefficient is negative, then the parabola opens downward.

The quantity $D = b^2 - 4ac$ is called the **discriminant** of the parabola. If the discriminant is positive, then the parabola has two real zeros. If the discriminant is negative, then it has no real zeros. If the discriminant is zero, then the parabola's highest or lowest point is on the x-axis, and it has a single real zero.

The highest or lowest point of the parabola is called the **vertex**. The coordinates of the vertex are given by the point $(-\frac{b}{2a}, -\frac{D}{4a})$. The roots of a quadratic function can be found with the quadratic formula, which is:

$$x = \frac{-b \pm \sqrt{b^2 - 4ac}}{2a}$$

A **rational function** is a function $f(x) = \frac{p(x)}{q(x)}$, where p and q are both polynomials. The domain of f will be all real numbers except the (real) roots of q. At these roots, the graph of f will have a **vertical asymptote,** unless they are also roots of p. Here is an example to consider:

$$p(x) = p_n x^n + p_{n-1} x^{n-1} + p_{n-2} x^{n-2} + \cdots + p_1 x + p_0$$

$$q(x) = q_m x^m + q_{m-1} x^{m-1} + q_{m-2} x^{m-2} + \cdots + q_1 x + q_0$$

When the degree of p is less than the degree of q, there will be a **horizontal asymptote** of $y = 0$. If p and q have the same degree, there will be a horizontal asymptote of $y = \frac{p_n}{q_n}$. If the degree of p is exactly one greater than the degree of q, then f will have an oblique asymptote along the line $y = \frac{p_n}{q_{n-1}} x + \frac{p_{n-1}}{q_{n-1}}$.

Exponential Functions

An **exponential function** is a function of the form $f(x) = b^x$, where b is a positive real number other than 1. In such a function, b is called the **base**.

The **domain** of an exponential function is all real numbers, and the **range** is all positive real numbers. There will always be a horizontal asymptote of $y = 0$ on one side. If b is greater than 1, then the graph will be increasing when moving to the right. If b is less than 1, then the graph will be decreasing when moving to the right. Exponential functions are one-to-one. The basic exponential function graph will go through the point (0, 1).

The following example demonstartes this more clearly:

Solve $5^{x+1} = 25$.

The first step is to get the x out of the exponent by rewriting the equation $5^{x+1} = 5^2$ so that both sides have a base of 5. Since the bases are the same, the exponents must be equal to each other. This leaves $x + 1 = 2$ or $x = 1$. To check the answer, the x-value of 1 can be substituted back into the original equation.

Logarithmic Functions

A **logarithmic function** is an inverse for an exponential function. The inverse of the base b exponential function is written as $\log_b(x)$, and is called the **base b logarithm**. The domain of a logarithm is all positive real numbers. It has the properties that $\log_b(b^x) = x$. For positive real values of x, $b^{\log_b(x)} = x$.

When there is no chance of confusion, the parentheses are sometimes skipped for logarithmic functions: $\log_b(x)$ may be written as $\log_b x$. For the special number e, the base e logarithm is called the **natural logarithm** and is written as $\ln x$. Logarithms are one-to-one.

When working with logarithmic functions, it is important to remember the following properties. Each one can be derived from the definition of the logarithm as the inverse to an exponential function:

- $\log_b 1 = 0$
- $\log_b b = 1$
- $\log_b b^p = p$
- $\log_b MN = \log_b M + \log_b N$
- $\log_b \frac{M}{N} = \log_b M - \log_b N$
- $\log_b M^p = p \log_b M$

When solving equations involving exponentials and logarithms, the following fact should be used:

If f is a one-to-one function, $a = b$ is equivalent to $f(a) = f(b)$.

Using this, together with the fact that logarithms and exponentials are inverses, allows for manipulations of the equations to isolate the variable as is demonstrated in the following example:

Solve $4 = \ln(x - 4)$.

Using the definition of a logarithm, the equation can be changed to $e^4 = e^{\ln(x-4)}$. The functions on the right side cancel with a result of $e^4 = x - 4$. This then gives $x = 4 + e^4$.

Trigonometric Functions

Trigonometric functions are built out of two basic functions, the **sine** and **cosine**, written as $\sin \theta$ and $\cos \theta$, respectively. Note that similar to logarithms, it is customary to drop the parentheses as long as the result is not confusing.

Sine and cosine are defined using the **unit circle**. If θ is the angle going counterclockwise around the origin from the x-axis, then the point on the unit circle in that direction will have the coordinates $(\cos \theta, \sin \theta)$.

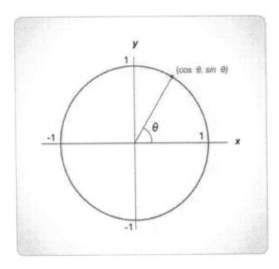

Since the angle returns to the start every 2π radians (or 360 degrees), the graph of these functions is **periodic**, with period 2π. This means that the graph repeats itself as one moves along the x-axis because $\sin \theta = \sin(\theta + 2\pi)$. Cosine works similarly.

From the unit circle definition, the sine function starts at 0 when $\theta = 0$. It grows to 1 as θ grows to $\pi/2$, and then back to 0 at $\theta = \pi$. Then it decreases to -1 as θ grows to $3\pi/2$, and goes back up to 0 at $\theta = 2\pi$.

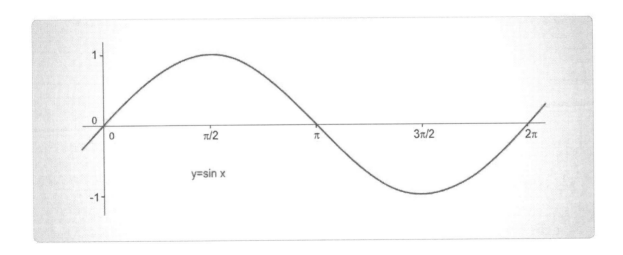

The graph of the cosine is similar. The cosine graph will start at 1, decreasing to 0 at $\pi/2$ and continuing to decrease to -1 at $\theta = \pi$. Then, it grows to 0 as θ grows to $3\pi/2$ and back up to 1 at $\theta = 2\pi$.

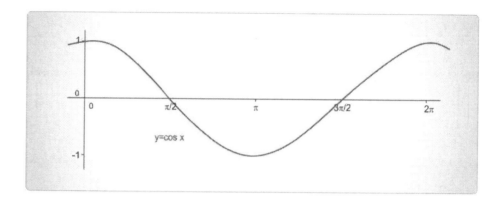

Another trigonometric function that is frequently used, is the **tangent** function. This is defined as the following equation: $\tan \theta = \dfrac{\sin \theta}{\cos \theta}$.

The tangent function is a period of π rather than 2π because the sine and cosine functions have the same absolute values after a change in the angle of π, but they flip their signs. Since the tangent is a ratio of the two functions, the changes in signs cancel.

The tangent function will be zero when sine is zero, and it will have a vertical asymptote whenever cosine is zero. The following graph shows the tangent function:

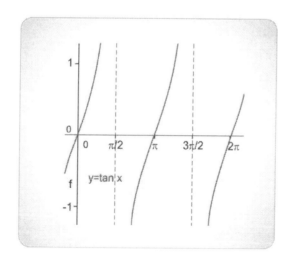

Three other trigonometric functions are sometimes useful. These are the **reciprocal** trigonometric functions, so named because they are just the reciprocals of sine, cosine, and tangent. They are the **cosecant**, defined as $\csc \theta = \dfrac{1}{\sin \theta}$, the **secant**, $\sec \theta = \dfrac{1}{\cos \theta}$, and the **cotangent**, $\cot \theta = \dfrac{1}{\tan \theta}$. Note that from the definition of tangent, $\cot \theta = \dfrac{\cos \theta}{\sin \theta}$.

In addition, there are three identities that relate the trigonometric functions to one another:

- $\cos \theta = \sin(\frac{\pi}{2} - \theta)$
- $\csc \theta = \sec\left(\frac{\pi}{2} - \theta\right)$
- $\cot \theta = \tan(\frac{\pi}{2} - \theta)$

Here is a list of commonly-needed values for trigonometric functions, given in radians, for the first quadrant:

Table for trigonometric functions

$\sin 0 = 0$	$\cos 0 = 1$	$\tan 0 = 0$
$\sin \frac{\pi}{6} = \frac{1}{2}$	$\cos \frac{\pi}{6} = \frac{\sqrt{3}}{2}$	$\tan \frac{\pi}{6} = \frac{\sqrt{3}}{3}$
$\sin \frac{\pi}{4} = \frac{\sqrt{2}}{2}$	$\cos \frac{\pi}{4} = \frac{\sqrt{2}}{2}$	$\tan \frac{\pi}{4} = 1$
$\sin \frac{\pi}{3} = \frac{\sqrt{3}}{2}$	$\cos \frac{\pi}{3} = \frac{1}{2}$	$\tan \frac{\pi}{3} = \sqrt{3}$
$\sin \frac{\pi}{2} = 1$	$\cos \frac{\pi}{2} = 0$	$\tan \frac{\pi}{2} = undefined$
$\csc 0 = undefined$	$\sec 0 = 1$	$\cot 0 = undefined$
$\csc \frac{\pi}{6} = 2$	$\sec \frac{\pi}{6} = \frac{2\sqrt{3}}{3}$	$\cot \frac{\pi}{6} = \sqrt{3}$
$\csc \frac{\pi}{4} = \sqrt{2}$	$\sec \frac{\pi}{4} = \sqrt{2}$	$\cot \frac{\pi}{4} = 1$
$\csc \frac{\pi}{3} = \frac{2\sqrt{3}}{3}$	$\sec \frac{\pi}{3} = 2$	$\cot \frac{\pi}{3} = \frac{\sqrt{3}}{3}$
$\csc \frac{\pi}{2} = 1$	$\sec \frac{\pi}{2} = undefined$	$\cot \frac{\pi}{2} = 0$

To find the trigonometric values in other quadrants, complementary angles can be used. The **complementary angle** is the smallest angle between the x-axis and the given angle.

Once the complementary angle is known, the following rule is used:

For an angle θ with complementary angle x, the absolute value of a trigonometric function evaluated at θ is the same as the absolute value when evaluated at x.

The correct sign for sine and cosine is determined by the x and y coordinates on the unit circle.

- ○ Sine will be positive in quadrants I and II and negative in quadrants III and IV.
- ○ Cosine will be positive in quadrants I and IV, and negative in II and III.
- ○ Tangent will be positive in I and III, and negative in II and IV.

The signs of the reciprocal functions will be the same as the sign of the function of which they are the reciprocal. For example:

Find $\sin\frac{3\pi}{4}$.

The complementary angle must be found first. This angle is in the II quadrant, and the angle between it and the x-axis is $\frac{\pi}{4}$. Now, $\sin\frac{\pi}{4} = \frac{\sqrt{2}}{2}$. Since this is in the II quadrant, sine takes on positive values (the y coordinate is positive in the II quadrant). Therefore, $\sin\frac{3\pi}{4} = \frac{\sqrt{2}}{2}$.

In addition to the six trigonometric functions defined above, there are inverses for these functions. However, since the trigonometric functions are not one-to-one, one can only construct inverses for them on a restricted domain.

Usually, the domain chosen will be $[0, \pi)$ for cosine and $(-\frac{\pi}{2}, \frac{\pi}{2}]$ for sine. The inverse for tangent can use either of these domains. The inverse functions for the trigonometric functions are also called **arc functions**. In addition to being written with a -1 as the exponent to denote that the function is an inverse, they will sometimes be written with an "a" or "arc" in front of the function name, so $\cos^{-1}\theta = a\cos\theta = \arccos\theta$.

When solving equations that involve trigonometric functions, there are often multiple solutions. For example, $2\sin\theta = \sqrt{2}$ can be simplified to $\sin\theta = \frac{\sqrt{2}}{2}$. This has solutions $\theta = \frac{\pi}{4}, \frac{3\pi}{4}$, but in addition, because of the periodicity, any integer multiple of 2π can also be added to these solutions to find another solution.

The full set of solutions is $\theta = \frac{\pi}{4} + 2\pi k, \frac{3\pi}{4} + 2\pi k$ for all integer values of k. It is very important to remember to find all possible solutions when dealing with equations that involve trigonometric functions.

The name *trigonometric* comes from the fact that these functions play an important role in the geometry of triangles, particularly right triangles. Consider the right triangle shown in this figure:

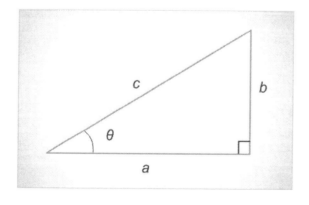

The following hold true:

- $c \sin \theta = b$.
- $c \cos \theta = a$.
- $\tan \theta = \dfrac{b}{a}$.
- $b \csc \theta = c$.
- $a \sec \theta = c$.
- $\cot \theta = \dfrac{a}{b}$.

It is important to remember that the angles of a triangle must add up to π radians (180 degrees).

Geometry

Shapes and Solids

Perimeter is the distance measurement around something. It can be thought of as the length of the boundary, like a fence. In contrast, area is the space occupied by a defined enclosure, like a field enclosed by a fence.

The perimeter of a square is measured by adding together all of the sides. Since a square has four equal sides, its perimeter can be calculated by multiplying the length of one side by 4. Thus, the formula is $P = 4 \times s$, where s equals one side. The area of a square is calculated by squaring the length of one side, which is expressed as the formula $A = s^2$.

Like a square, a rectangle's perimeter is measured by adding together all of the sides. But as the sides are unequal, the formula is different. A rectangle has equal values for its lengths (long sides) and equal values for its widths (short sides), so the perimeter formula for a rectangle is $P = l + l + w + w = 2l + 2w$, where l equals length and w equals width. The area is found by multiplying the length by the width, so the formula is $A = l \times w$.

A triangle's perimeter is measured by adding together the three sides, so the formula is $P = a + b + c$, where $a, b,$ and c are the values of the three sides. The area is calculated by multiplying the length of the base times the height times ½, so the formula is $A = \frac{1}{2} \times b \times h = \frac{bh}{2}$. The base is the bottom of the triangle, and the height is the distance from the base to the peak. If a problem asks one to calculate the area of a triangle, it will provide the base and height.

A circle's perimeter—also known as its **circumference**—is measured by multiplying the **diameter** (the straight line measured from one side, through the center, to the direct opposite side of the circle) by π, so the formula is $\pi \times d$. This is sometimes expressed by the formula $C = 2 \times \pi \times r$, where r is the **radius** of the circle. These formulas are equivalent, as the radius equals half of the diameter. The area of a circle is calculated with the formula $A = \pi \times r^2$. The test will indicate either to leave the answer with π attached or to calculate to the nearest decimal place, which means multiplying by 3.14 for π.

The perimeter of a parallelogram is measured by adding the lengths and widths together. Thus, the formula is the same as for a rectangle, $P = l + l + w + w = 2l + 2w$. However, the area formula differs from the rectangle. For a parallelogram, the area is calculated by multiplying the length by the height: $A = h \times l$

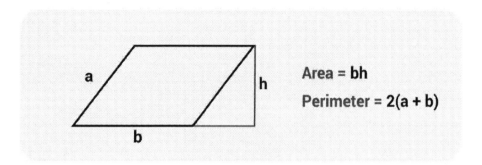

The perimeter of a trapezoid is calculated by adding the two unequal bases and two equal sides, so the formula is $P = a + b_1 + c + b_2$. Although unlikely to be a test question, the formula for the area of a trapezoid is $A = \frac{b_1 + b_2}{2} \times h$, where h equals height, and b_1 and b_2 equal the bases.

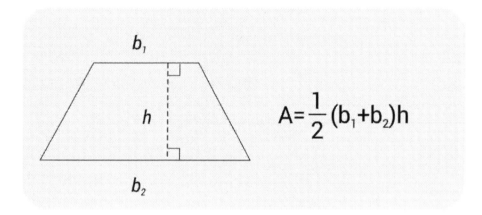

Congruence and Similarity

Triangles are similar if they have the same angle measurements, and their sides are proportional to one another. Triangles are **congruent** if the angles of the triangles are equal in measurement and the sides of the triangles are equal in measurement.

There are five ways to show that triangles are congruent:

1. SSS (Side-Side-Side Postulate) – when all three corresponding sides are equal in length, then the two triangles are congruent.

2. SAS (Side-Angle-Side Postulate) – if a pair of corresponding sides and the angle in between those two sides are equal, then the two triangles are congruent.

3. ASA (Angle-Side-Angle Postulate) – if a pair of corresponding angles are equal and the side lengths within those angles are equal, then the two triangles are equal.

4. AAS (Angle-Angle-Side Postulate) – when a pair of corresponding angles for two triangles and a non-included side are equal, then the two triangles are congruent.

5. HL (Hypotenuse-Leg Theorem) – if two right triangles have the same hypotenuse length, and one of the other sides in each triangle are of the same length, then the two triangles are congruent.

If two triangles are discovered to be similar or congruent, this information can assist in determining unknown parts of triangles, such as missing angles and sides.

The example below involves the question of congruent triangles. The first step is to examine whether the triangles are congruent. If the triangles are congruent, then the measure of a missing angle can be found.

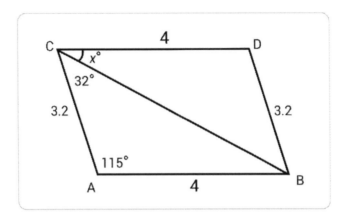

The above diagram provides values for angle measurements and side lengths in triangles *CAB* and *CDB*. Note that side *CA* is 3.2 and side *DB* is 3.2. Side *CD* is 4 and side *AB* is 4. Furthermore, line *CB* is congruent to itself by the reflexive property. Therefore, the two triangles are congruent by SSS (Side-Side-Side). Because the two triangles are congruent, all of the corresponding parts of the triangles are also congruent. Therefore, angle *x* is congruent to the inside of the angle for which a measurement is not provided in triangle *CAB*. Thus, 115º + 32º = 147º. A triangle's angles sum to 180º, therefore, 180º – 147º = 33º. Angle *x* = 33º, because the two triangles are reversed.

Surface Area and Volume

Surface area and volume are two- and three-dimensional measurements. Surface area measures the total surface space of an object, like the six sides of a cube. Questions about surface area will ask how much of something is needed to cover a three-dimensional object, like wrapping a present. **Volume** is the measurement of how much space an object occupies, like how much space is in the cube. Volume questions will ask how much of something is needed to completely fill the object. The most common surface area and volume questions deal with spheres, cubes, and rectangular prisms.

The formula for a cube's surface area is $SA = 6 \times s^2$, where s is the length of a side. A cube has 6 equal sides, so the formula expresses the area of all the sides. Volume is simply measured by taking the cube of the length, so the formula is $V = s^3$.

The surface area formula for a rectangular prism or a general box is SA = $2(lw + lh + wh)$, where l is the length, h is the height, and w is the width. The volume formula is $V = l \times w \times h$, which is the cube's volume formula adjusted for the unequal lengths of a box's sides.

The formula for a sphere's surface area is $SA = 4\pi r^2$, where r is the sphere's radius. The surface area formula is the area for a circle multiplied by four. To measure volume, the formula is $V = \frac{4}{3}\pi r^3$.

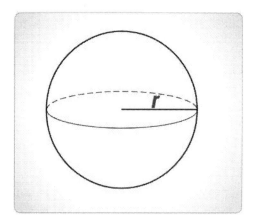

A **rectangular pyramid** is a figure with a rectangular base and four triangular sides that meet at a single vertex. If the rectangle has sides of lengths x and y, then the volume will be given by $V = \frac{1}{3}xyh$.

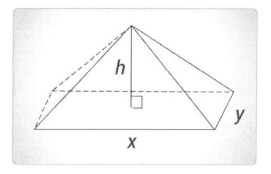

To find the surface area, the dimensions of each triangle must be known. However, these dimensions can differ depending on the problem in question. Therefore, there is no general formula for calculating total surface area.

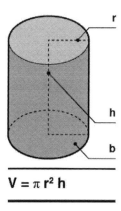

$$V = \pi r^2 h$$

The formula to find the volume of a cylinder is $\pi r^2 h$. This formula contains the formula for the area of a circle (πr^2) because the base of a cylinder is a circle. To calculate the volume of a cylinder, the slices of circles needed to build the entire height of the cylinder are added together. For example, if the radius is 5 feet and the height of the cylinder is 10 feet, the cylinder's volume is calculated by using the following equation: $\pi 5^2 \times 10$. Substituting 3.14 for π, the volume is 785.4 ft³.

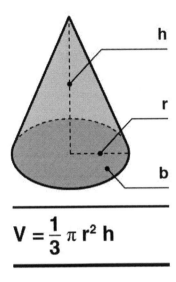

$$V = \frac{1}{3} \pi \, r^2 \, h$$

The formula used to calculate the volume of a cone is $\frac{1}{3}\pi r^2 h$. Essentially, the area of the base of the cone is multiplied by the cone's height. In a real-life example where the radius of a cone is 2 meters and the height of a cone is 5 meters, the volume of the cone is calculated by utilizing the formula $\frac{1}{3}\pi 2^2 \times 5 = 21$. After substituting 3.14 for π, the volume is 785.4 ft³.

Solving for Missing Values in Triangles

Suppose that Lara is 5 feet tall and is standing 30 feet from the base of a light pole, and her shadow is 6 feet long. How high is the light on the pole? To figure this out, it helps to make a sketch of the situation:

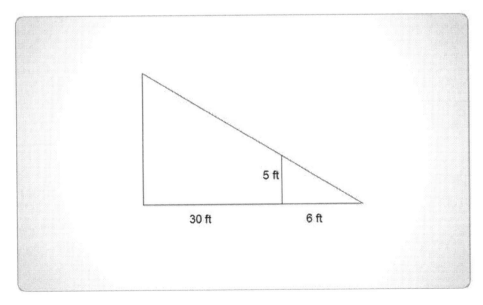

The light pole is the left side of the triangle. Lara is the 5-foot vertical line. Test takers should notice that there are two right triangles here, and that they have all the same angles as one another. Therefore, they form similar triangles. So, the ratio of proportionality between them must be found.

The bases of these triangles are known. The small triangle, formed by Lara and her shadow, has a base of 6 feet. The large triangle, formed by the light pole along with the line from the base of the pole out to the end of Lara's shadow is $30 + 6 = 36$ feet long. So, the ratio of the big triangle to the little triangle is $\frac{36}{6} = 6$. The height of the little triangle is 5 feet. Therefore, the height of the big triangle will be $6 \cdot 5 = 30$ feet, meaning that the light is 30 feet up the pole.

The Pythagorean Theorem

The **Pythagorean theorem** states that for right triangles, the sum of the squares of the two shorter sides is equal to the square of the longest side (also called the **hypotenuse**). The longest side will always be the side opposite to the 90° angle. If this side is called c, and the other two sides are a and b, then the Pythagorean theorem states that $c^2 = a^2 + b^2$. Since lengths are always positive, this also can be written as $c = \sqrt{a^2 + b^2}$. A diagram to show the parts of a triangle using the Pythagorean theorem is below.

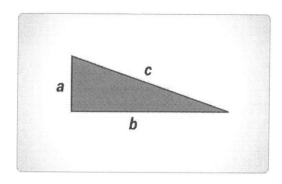

As an example of the theorem, suppose that Shirley has a rectangular field that is 5 feet wide and 12 feet long, and she wants to split it in half using a fence that goes from one corner to the opposite corner. How long will this fence need to be? To figure this out, note that this makes the field into two right triangles, whose hypotenuse will be the fence dividing it in half. Therefore, the fence length is given by $\sqrt{5^2 + 12^2} = \sqrt{169} = 13$ feet long.

One last useful relationship between the trigonometric functions introduced in the *Functions* section is the **Pythagorean identity**, which states that $\sin^2 \theta + \cos^2 \theta = 1$. Note that for trigonometric functions, the exponent is sometimes written next to the function name, so $\sin^2 \theta = (\sin \theta)^2$, and so on. The same is sometimes also done for logarithmic functions. The Pythagorean identity has two other forms which are often useful: $1 + \cot^2 \theta = \csc^2 \theta$ and $\tan^2 \theta + 1 = \sec^2 \theta$.

As mentioned, although the trigonometric functions are not one-to-one, it is possible to define inverses for them on limited domains. These are also called the *arc* functions. The inverse function for sine is called the **arcsine** and is written as either $\sin^{-1} x$ or as $\mathrm{asin}\, x$, and similarly for the other five trigonometric functions. The range of the arcsine and arccosecant is usually taken to be $[-\frac{\pi}{2}, \frac{\pi}{2}]$. The range of the arccosine, arcsecant, arctangent, and arccotangent are generally taken to be $[0, \pi]$. Some specific values for these inverse functions can be read off published tables.

When solving an equation using these inverses, unless the domain is specifically restricted, all possible angles which satisfy the equation must be considered. For example, when solving $\cos(x - 1) = \frac{\sqrt{2}}{2}$, arccosine is applied to both sides, which yields $x - 1 = a\cos\frac{\sqrt{2}}{2}$. From the available tables, cosine takes the value $\frac{\sqrt{2}}{2}$ for any angle $2\pi k \pm \frac{\pi}{4}$, where k is an arbitrary integer. The equation becomes $x - 1 = 2\pi k \pm \frac{\pi}{4}$, or $x = 2\pi k \pm \frac{\pi}{4} + 1$.

Performing Algebraic Operations on Functions

As mentioned, it is possible to perform algebraic operations between functions, meaning they can be added, subtracted, multiplied, or divided. In fact, all the trigonometric functions are formed this way from sine and cosine. More generally, everything stated regarding arithmetic operations on functions can be done for trigonometric and logarithmic functions. However, sometimes it will be possible to use their definition to simplify the result.

For example, given $f(x) = \sin x + 1, g(x) = \cos x$, find $\frac{f(x)}{g(x)}$. This will, of course, be $\frac{\sin x + 1}{\cos x}$; however, this can be further simplified using the identities of trigonometric functions. The expression can be rewritten as $\frac{\sin x}{\cos x} + \frac{1}{\cos x} = \tan x + \sec x$.

Identifying and Using Composite Functions

Everything previously explained about composing functions can be applied to exponential, logarithmic, and trigonometric functions as well. For example, given $f(x) = 5^x, g(x) = \cos x$, one may form the composition $(g \circ f)(x) = \cos(5^x)$. The ability to recognize such compositions will be particularly important when discussing calculus, where more examples will be considered.

Conics

The graph of an equation of the form $y = ax^2 + bx + c$ or $x = ay^2 + by + c$ is called a **parabola.**

The graph of an equation of the form $\frac{x^2}{a^2} - \frac{y^2}{b^2} = 1$ or $-\frac{x^2}{a^2} + \frac{y^2}{b^2} = 1$ is called a **hyperbola**.

The graph of an equation of the form $\frac{(x-x_0)^2}{a^2} + \frac{(y-y_0)^2}{b^2} = 1$ is called an **ellipse**. If $a = b$ then this is a circle with **radius** $r = \frac{1}{a}$.

Statistics and Probability

Center and Spread of Distributions

Descriptive statistics are utilized to gain an understanding of properties of a data set. This entails examining the center, spread, and shape of the sample data.

Center
The **center** of the sample set can be represented by its mean, median or mode. The **mean** is the average of the data set. It is calculated by adding the data values together and dividing this sum by the sample size (the number of data points). The **median** is the value of the data point in the middle when the sample is arranged in numerical order. If the sample has an even number of data points, the mean of the

two middle values is the median. The **mode** is the value which appears most often in a data set. It is possible to have multiple modes (if different values repeat equally as often) or no mode (if no value repeats).

Spread

Methods for determining the **spread** of the sample include calculating the range and standard deviation for the data. The *range* is calculated by subtracting the lowest value from the highest value in the set. The **standard deviation** of the sample can be calculated using the formula:

$$\sigma = \sqrt{\frac{\sum(x - \bar{x})^2}{n - 1}}$$

$\bar{x}$ = sample mean
n = sample size

Shape

The **shape** of the sample when displayed as a histogram or frequency distribution plot helps to determine if the sample is normally distributed (bell-shaped curve), symmetrical, or displays skewness (lack of symmetry), or kurtosis. **Kurtosis** is a measure of whether the data are heavy-tailed (high number of outliers) or light-tailed (low number of outliers).

Data Collection Methods

Statistical inference, based in probability theory, makes calculated assumptions about an entire population based on data from a sample set from that population.

Population Parameters

A population is the entire set of people or things of interest. For example, if researchers wanted to determine the number of hours of sleep per night for college females in the U.S, the population would consist of *every* college female in the country. A **sample** is a subset of the population that may be used for the study. A sample might consist of 100 students per school from 20 different colleges in the country. From the results of the survey, a sample statistic can be calculated. A **sample statistic** is a numerical characteristic of the sample data including mean and variance. A sample statistic can be used to estimate a corresponding **population parameter**, which is a numerical characteristic of the entire population.

Confidence Intervals

A population parameter estimated using a sample statistic may be very accurate or relatively inaccurate based on errors in sampling. A **confidence interval** indicates a range of values likely to include the true population parameter. A given confidence interval such as 95% means that the true population parameter will occur within the interval for 95% of samples.

Measurement Error

The accuracy of a population parameter based on a sample statistic may also be affected by measurement error. **Measurement error** can be divided into random error and systematic error. An example of **random error** for the previous scenario would be a student reporting 8 hours of sleep when she actually sleeps 7 hours per night. **Systematic errors** are those attributed to the measurement system. If the sleep survey gave response options of 2,4,6,8, or 10 hours. This would lead to systematic measurement error because certain values could not be accurately reported.

Evaluating Reports and Determining the Appropriateness of Data Collection Methods

The presentation of statistics can be manipulated to produce a desired outcome. For example, in the statement "four out of five dentists recommend our toothpaste", critical readers should wonder: *who are the five dentists?* While the wording is similar, this statement is very different from "four out of every five dentists recommend our toothpaste." The context of the numerical values allows one to decipher the meaning, intent, and significance of the survey or study.

When analyzing a report, the researchers who conducted the study and their intent must be considered. Was it performed by a neutral party or by a person or group with a vested interest? The sampling method and the data collection method should also be evaluated. Was it a true random sample of the population or was one subgroup over- or underrepresented? Lastly, the measurement system used to obtain the data should be assessed. Was the system accurate and precise or was it a flawed system?

Understanding and Modeling Relationships in Bivariate Data

The simplest type of correlation between two variables is a **linear correlation**. If the independent variable is x and the dependent variable is y, then a linear correlation means $y = mx + b$. If m is positive, then y will increase as x increases. While if m is negative, then y decreases while x increases. The variable b represents the value of y when x is 0.

Calculating Probabilities, Including Related Sample Spaces

Probability, represented by variable p, always has a value from 0 to 1. The total probability for all the possible outcomes (sample space) should equal 1.

The probability of a single outcome x_i can be expressed $p(x_i) = \frac{1}{n}$, where n is the total number of possible outcomes. A good example of this is a fair six-sided die, in which the possible outcomes are 1, 2, 3, 4, 5, and 6, and the individual probability of each of these six outcomes is $\frac{1}{6}$.

The probability of an outcome occurring from a range A of possible outcomes is written as $P(A)$. To compute this, the probabilities for each outcome in A are added together. To use the example of a fair six-sided die, a problem may ask one to find the probability of getting a 2 or lower when it is rolled. The possible rolls are 1, 2, 3, 4, 5, and 6. So, to get a 2 or lower, one must roll a 1 or a 2. Each probability is $\frac{1}{6}$, and adding them together to get $p(1) + p(2) = \frac{1}{6} + \frac{1}{6} = \frac{1}{3}$.

Here are a few types of probability distributions that are standard and have standard names:

- The **binomial distribution**: This distribution describes the probability of getting k successes in n trials, where each trial can either succeed or fail. If the probability of a single trial being a success is denoted by p, then, the probability of getting k successes in n trials is:

$$\frac{n!}{k!\,(n-k)!}p^k(1-p)^{n-k}$$

- A **Poisson distribution**: This describes the probability of getting k events in a fixed interval of time. If the average number of events during this time interval is denoted by λ, then the probability of getting k events during this time interval is given by $\frac{\lambda^k e^{-\lambda}}{k!}$.

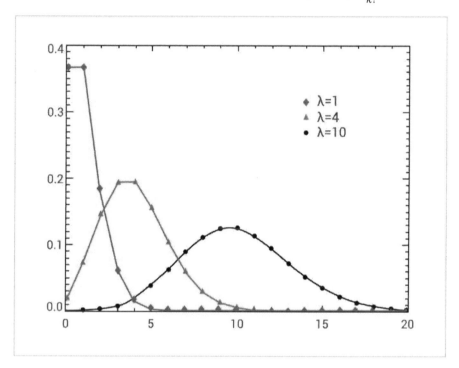

Conditional Probabilities

In some cases, it may be known that the outcome lies within the subset of possibilities B, and a problem will ask for probability that the outcome will end up being inside another subset of possibilities A. This kind of problem is called a **conditional probability**.

As an application of this, one can imagine a fair die is rolled. It is known that the roll's value lies between 1 and 4, inclusive, but the number of sides that the die has is not known (not all dice are six-sided). A problem may ask for the probability the roll was higher than 2. In this case, the total number of sides of the die is unimportant, and, since this is a fair die, the probability distribution is uniform across all possibilities. Therefore, it is possible to apply the formula $\frac{|A \cap B|}{|B|}$. $A \cap B$ means "A intersect B," and is the set of all outcomes that lie in both A and B. In this particular problem, B is {1, 2, 3, 4} and A ∩ B is {3, 4}. Therefore, $\frac{|A \cap B|}{|B|} = \frac{2}{4} = \frac{1}{2}$.

In many cases, changing the order of the conditional probabilities will greatly affect the outcome. For example, if a person has received a military medal, it is certain the person must have served in the military. However, given that a person served in the military, the probability of them receiving a military medal may not be very high.

In some special cases, however, the order in which the conditional probabilities are taken will not change the final probability. In this case, A and B are said to be **independent**. One situation in which one would expect that the outcomes should be independent is in rolling a pair of dice. If someone rolls two dice (a white die and a black die), then one would expect that the number that is obtained from the white die should not depend upon the number that is obtained from the black die, or vice versa. The same principle applies to a situation in which one rolls a single die repeatedly. A similar situation applies to flipping a coin repeatedly: whether the next flip is heads or tails will not depend upon what results have been obtained in previous flips.

ACT Math Practice Test #1

1. At the beginning of the day, Xavier has 20 apples. At lunch, he meets his sister Emma and gives her half of his apples. After lunch, he stops by his neighbor Jim's house and gives him 6 of his apples. He then uses ¾ of his remaining apples to make an apple pie for dessert at dinner. At the end of the day, how many apples does Xavier have left?

    a. 4
    b. 6
    c. 2
    d. 1
    e. 3

2. What is the product of two irrational numbers?

    a. Irrational
    b. Rational
    c. Irrational or rational
    d. Complex and imaginary
    e. Imaginary

3. The graph shows the position of a car over a 10-second time interval. Which of the following is the correct interpretation of the graph for the interval 1 to 3 seconds?

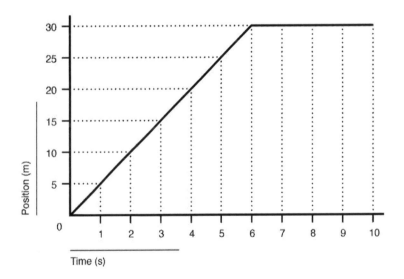

    a. The car remains in the same position.
    b. The car is traveling at a speed of 5m/s.
    c. The car is traveling up a hill.
    d. The car is traveling at 5mph.
    e. The car accelerates at a rate of 5m/s.

4. Being as specific as possible, how is the number -4 classified?
   a. Real, rational, integer, whole, natural
   b. Real, rational, integer, natural
   c. Real, rational, integer
   d. Real, irrational, complex
   e. Real, irrational, whole

5. After a 20% discount, Frank purchased a new refrigerator for $850. How much did he save from the original price?
   a. $170
   b. $212.50
   c. $105.75
   d. $200
   e. $187.50

6. $5.88 \times 3.2 =$
   a. 15.816
   b. 16.44
   c. 20.352
   d. 17
   e. 18.816

7. What is the solution to the following problem in decimal form?
$$\frac{3}{5} \times \frac{7}{10} \div \frac{1}{2}$$
   a. 0.042
   b. 84%
   c. 0.84
   d. 0.42
   e. 42%

8. Dwayne has received the following scores on his math tests: 78, 92, 83, and 97. What score must Dwayne get on his next math test to have an overall average of at least 90?
   a. 89
   b. 98
   c. 95
   d. 94
   e. 100

9. What are all the factors of 12?
   a. 12, 24, 36
   b. 1, 2, 4, 6, 12
   c. 12, 24, 36, 48
   d. 1, 2, 3, 4, 6, 12
   e. 0, 1, 12

10. Which of the following augmented matrices represents the system of equations below?

$$2x - 3y + z = -5$$
$$4x - y - 2z = -7$$
$$-x + 2z = -1$$

a. $\begin{bmatrix} 2 & -3 & 1 & 5 \\ 4 & -1 & 0 & -7 \\ -1 & 0 & 2 & 1 \end{bmatrix}$

b. $\begin{bmatrix} 2 & 4 & -1 \\ -3 & -1 & 0 \\ 1 & -2 & 2 \\ -5 & -7 & -1 \end{bmatrix}$

c. $\begin{bmatrix} 2 & 4 & -1 & -5 \\ -3 & -1 & 0 & -7 \\ 2 & -2 & 2 & -1 \end{bmatrix}$

d. $\begin{bmatrix} 2 & -3 & 1 \\ 4 & -1 & -2 \\ -1 & 0 & 2 \end{bmatrix}$

e. $\begin{bmatrix} 2 & -3 & 1 & -5 \\ 4 & -1 & -2 & -7 \\ -1 & 0 & 2 & -1 \end{bmatrix}$

11. What are the zeros of the function: $f(x) = x^3 + 4x^2 + 4x$?
a. -2
b. 0, -2
c. 2
d. 0, 2
e. 0

12. If $g(x) = x^3 - 3x^2 - 2x + 6$ and $f(x) = 2$, then what is $g(f(x))$?
a. -26
b. 6
c. $2x^3 - 6x^2 - 4x + 12$
d. -2
e. $2x^2 - 6$

13. What is the solution to the following system of equations?

$$x^2 - 2x + y = 8$$
$$x - y = -2$$

a. $(-2, 3)$
b. There is no solution.
c. $(-2, 0) \ (1, 3)$
d. $(-2, 0) \ (3, 5)$
e. $(2, 0) \ (-1, 3)$

14. Which of the following is the result after simplifying the expression: $(7n + 3n^3 + 3) + (8n + 5n^3 + 2n^4)$?

    a. $9n^4 + 15n - 2$
    b. $2n^4 + 5n^3 + 15n - 2$
    c. $9n^4 + 8n^3 + 15n$
    d. $2n^4 + 8n^3 + 15n + 3$
    e. $2n^4 + 5n^3 + 15n - 3$

15. What is the product of the following expression?
$$(4x - 8)(5x^2 + x + 6)$$

    a. $20x^3 - 36x^2 + 16x - 48$
    b. $6x^3 - 41x^2 + 12x + 15$
    c. $204 + 11x^2 - 37x - 12$
    d. $2x^3 - 11x^2 - 32x + 20$
    e. $20x^3 - 40x^2 + 24x - 48$

16. How could the following equation be factored to find the zeros?
$$y = x^3 - 3x^2 - 4x$$

    a. $0 = x^2(x - 4), x = 0, 4$
    b. $0 = 3x(x + 1)(x + 4), x = 0, -1, -4$
    c. $0 = x(x + 1)(x + 6), x = 0, -1, -6$
    d. $0 = 3x(x + 1)(x - 4), x = 0, 1, -4$
    e. $0 = x(x + 1)(x - 4), x = 0, -1, 4$

17. How will the number 847.89632 be written if rounded to the nearest hundredth?

    a. 847.90
    b. 900
    c. 847.89
    d. 847.896
    e. 847.9

18. Which of the following is the solution for the given equation?
$$\frac{x^2 + x - 30}{x - 5} = 11$$

    a. $x = -6$
    b. All real numbers.
    c. $x = 16$
    d. $x = 5$
    e. There is no solution.

19. Mom's car drove 72 miles in 90 minutes. How fast did she drive in feet per second?
    a. 0.8 feet per second
    b. 48.9 feet per second
    c. 0.009 feet per second
    d. 70.4 feet per second
    e. 21.3 feet per second

20. Solve $V = lwh$ for h.
    a. $lwV = h$
    b. $h = \dfrac{V}{lw}$
    c. $h = \dfrac{Vl}{w}$
    d. $h = \dfrac{Vw}{l}$
    e. $h = \dfrac{Vl}{w}$

21. What is the domain for the function $y = \sqrt{x}$?
    a. All real numbers
    b. $x \geq 0$
    c. $x > 0$
    d. $y \geq 0$
    e. $x < 0$

22. If Sarah reads at an average rate of 21 pages in four nights, how long will it take her to read 140 pages?
    a. 6 nights
    b. 26 nights
    c. 8 nights
    d. 27 nights
    e. 21 nights

23. The phone bill is calculated each month using the equation $c = 50g + 75$. The cost of the phone bill per month is represented by c, and g represents the gigabytes of data used that month. Identify and interpret the slope of this equation.
    a. 75 dollars per day
    b. 75 gigabytes per day
    c. 50 dollars per day
    d. 50 dollars per gigabyte
    e. The slope cannot be determined

24. What is the function that forms an equivalent graph to $y = \cos(x)$?
    a. $y = \tan(x)$
    b. $y = \csc(x)$
    c. $y = \sin\left(x + \dfrac{\pi}{2}\right)$
    d. $y = \sin\left(x - \dfrac{\pi}{2}\right)$
    e. $y = \tan\left(x + \dfrac{\pi}{2}\right)$

25. If $\sqrt{1+x} = 4$, what is x?

    a. 10

    b. 15

    c. 20

    d. 25

    e. 36

26. What is the inverse of the function $f(x) = 3x - 5$?

    a. $f^{-1}(x) = \frac{x}{3} + 5$

    b. $f^{-1}(x) = \frac{5x}{3}$

    c. $f^{-1}(x) = 3x + 5$

    d. $f^{-1}(x) = \frac{x+5}{3}$

    e. $f^{-1}(x) = \frac{x}{3} - 5$

27. What are the zeros of $f(x) = x^2 + 4$?

    a. $x = -4$

    b. $x = \pm 2i$

    c. $x = \pm 2$

    d. $x = \pm 4i$

    e. $x = 2, 4$

28. Twenty is 40% of what number?

    a. 60

    b. 8

    c. 200

    d. 70

    e. 50

29. What is the simplified form of the expression $1.2 * 10^{12} \div 3.0 * 10^8$?

    a. $0.4 * 10^4$

    b. $4.0 * 10^4$

    c. $4.0 * 10^3$

    d. $3.6 * 10^{20}$

    e. $4.0 * 10^2$

30. You measure the width of your door to be 36 inches. The true width of the door is 35.75 inches. What is the relative error in your measurement?

    a. 0.7%

    b. 0.007%

    c. 0.99%

    d. 0.1%

    e. 7.0%

31. What are the y-intercept(s) for $y = x^2 + 3x - 4$?
    a. $y = 1$
    b. $y = -4$
    c. $y = 3$
    d. $y = 4$
    e. $y = -3$

32. Is the following function even, odd, neither, or both?
$$y = \frac{1}{2}x^4 + 2x^2 - 6$$
    a. Even
    b. Odd
    c. Neither
    d. Both
    e. Even for all negative x-values and odd for all positive x-values

33. Which equation is not a function?
    a. $y = |x|$
    b. $y = \sqrt{x}$
    c. $x = 3$
    d. $y = 4$
    e. $y = 3x$

34. How could the following function be rewritten to identify the zeros?
$$y = 3x^3 + 3x^2 - 18x$$
    a. $y = 3x(x + 3)(x - 2)$
    b. $y = x(x - 2)(x + 3)$
    c. $y = 3x(x - 3)(x + 2)$
    d. $y = (x + 3)(x - 2)$
    e. $y = 3x(x + 3)(x + 2)$

35. A six-sided die is rolled. What is the probability that the roll is 1 or 2?
    a. $\frac{1}{6}$
    b. $\frac{1}{4}$
    c. $\frac{1}{3}$
    d. $\frac{1}{2}$
    e. $\frac{1}{36}$

36. A line passes through the origin and through the point (-3, 4). What is the slope of the line?
    a. $-\frac{4}{3}$
    b. $-\frac{3}{4}$
    c. $\frac{4}{3}$
    d. $\frac{3}{4}$
    e. $\frac{1}{3}$

37. What type of function is modeled by the values in the following table?

x	$f(x)$
1	2
2	4
3	8
4	16
5	32

    a. Linear
    b. Exponential
    c. Quadratic
    d. Cubic
    e. Logarithmic

38. An investment of $2,000 is made into an account with an annual interest rate of 5%, compounded continuously. What is the total value of the investment after eight years?
    a. $4,707
    b. $3,000
    c. $2,983.65
    d. $10,919.63
    e. $1,977.61

39. A ball is drawn at random from a ball pit containing 8 red balls, 7 yellow balls, 6 green balls, and 5 purple balls. What's the probability that the ball drawn is yellow?
    a. $1/26$
    b. $19/26$
    c. $14/26$
    d. 1
    e. $7/26$

40. Two cards are drawn from a shuffled deck of 52 cards. What's the probability that both cards are Kings if the first card isn't replaced after it's drawn and is a King?
    a. $1/169$
    b. $1/221$
    c. $1/13$
    d. $4/13$
    e. $1/104$

41. What's the probability of rolling a 6 at least once in two rolls of a die?

a. $1/3$

b. $1/36$

c. $1/6$

d. $1/12$

e. $11/36$

42. Given the set $A = \{1, 2, 3, 4, 5, 6, 7, 8, 9, 10\}$ and $B = \{1, 2, 3, 4, 5\}$, what is $A - (A \cap B)$?

a. $\{6, 7, 8, 9, 10\}$

b. $\{1, 2, 3, 4, 5\}$

c. $\{1, 2, 3, 4, 5, 6, 7, 8, 9, 10\}$

d. $\emptyset$

e. $\{-1, -2, -3, -4, -5\}$

43. An equilateral triangle has a perimeter of 18 feet. If a square whose sides have the same length as one side of the triangle is built, what will be the area of the square?

a. 6 square feet

b. 36 square feet

c. 256 square feet

d. 1000 square feet

e. 324 square feet

44. In a group of 20 men, the median weight is 180 pounds and the range is 30 pounds. If each man gains 10 pounds, which of the following would be true?

a. The median weight will increase, and the range will remain the same.

b. The median weight and range will both remain the same.

c. The median weight will stay the same, and the range will increase.

d. The median weight and range will both increase.

e. The median weight will increase, and the range will decrease.

45. For the following similar triangles, what are the values of x and y (rounded to one decimal place)?

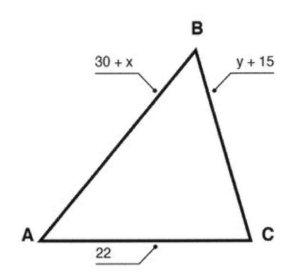 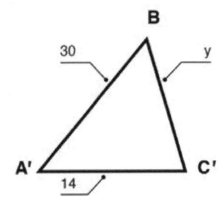

a. $x = 16.5, y = 25.1$
b. $x = 19.5, y = 24.1$
c. $x = 17.1, y = 26.3$
d. $x = 26.3, y = 17.1$
e. $x = 24.1, y = 19.5$

46. On Monday, Robert mopped the floor in 4 hours. On Tuesday, he did it in 3 hours. If on Monday, his average rate of mopping was p sq. ft. per hour, what was his average rate on Tuesday?

a. $\frac{4}{3}p$ sq. ft. per hour
b. $\frac{3}{4}p$ sq. ft. per hour
c. $\frac{5}{4}p$ sq. ft. per hour
d. $p + 1$ sq. ft. per hour
e. $\frac{1}{3}p$ sq. ft. per hour

47. Which of the following inequalities is equivalent to $3 - \frac{1}{2}x \geq 2$?

a. $x \geq 2$
b. $x \leq 2$
c. $x \geq 1$
d. $x \leq 1$
e. $x \leq -2$

48. For which of the following are $x = 4$ and $x = -4$ solutions?

a. $x^2 + 16 = 0$
b. $x^2 + 4x - 4 = 0$
c. $x^2 - 2x - 2 = 0$
d. $x^2 - x - 16 = 0$
e. $x^2 - 16 = 0$

49. What are the center and radius of a circle with equation $4x^2 + 4y^2 - 16x - 24y + 51 = 0$?
    a. Center (3, 2) and radius ½
    b. Center (2, 3) and radius ½
    c. Center (3, 2) and radius ¼
    d. Center (2, 3) and radius ¼
    e. Center (2, 2) and radius ¼

50. If the ordered pair $(-3, -4)$ is reflected over the x-axis, what's the new ordered pair?
    a. $(-3, -4)$
    b. $(3, -4)$
    c. $(3, 4)$
    d. $(-3, 4)$
    e. $(-4, -3)$

51. If the volume of a sphere is 288π cubic meters, what are the radius and surface area of the same sphere?
    a. Radius is 6 meters and surface area is 144π square meters
    b. Radius is 36 meters and surface area is 144π square meters
    c. Radius is 6 meters and surface area is 12π square meters
    d. Radius is 36 meters and surface area is 12π square meters
    e. Radius 12 meters and surface area 144π square meters

52. The triangle shown below is a right triangle. What's the value of x?

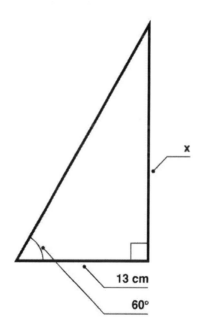

13 cm

60°

    a. $x = 1.73$
    b. $x = 0.57$
    c. $x = 13$
    d. $x = 14.73$
    e. $x = 22.49$

53. Ten students take a test. Five students get a 50. Four students get a 70. If the average score is 55, what was the last student's score?

    a. 20

    b. 40

    c. 50

    d. 60

    e. 62

54. A sample data set contains the following values: 1, 3, 5, 7. What's the standard deviation of the set?

    a. 2.58

    b. 4

    c. 6.23

    d. 1.1

    e. 0.25

55. A company invests $50,000 in a building where they can produce saws. If the cost of producing one saw is $40, then which function expresses the amount of money the company pays? The variable y is the money paid and x is the number of saws produced.

    a. $y = 50{,}000x + 40$

    b. $y + 40 = x - 50{,}000$

    c. $y = 40x - 50{,}000$

    d. $y = 40x + 50{,}000$

    e. $y = 4x + 50{,}000$

56. A pair of dice is thrown, and the sum of the two scores is calculated. What's the expected value of the roll?

    a. 5

    b. 6

    c. 7

    d. 8

    e. 9

57. A line passes through the point (1, 2) and crosses the y-axis at $y = 1$. Which of the following is an equation for this line?

    a. $y = 2x$

    b. $y = x + 1$

    c. $x + y = 1$

    d. $y = \frac{x}{2} - 2$

    e. $y = x - 1$

58. $x^4 - 16$ can be simplified to which of the following?

    a. $(x^2 - 4)(x^2 + 4)$

    b. $(x^2 + 4)(x^2 + 4)$

    c. $(x^2 - 4)(x^2 - 4)$

    d. $(x^2 - 2)(x^2 + 4)$

    e. $(x^2 - 2)(x^2 + 2)$

59. $(4x^2y^4)^{\frac{3}{2}}$ can be simplified to which of the following?

    a. $8x^3y^6$

    b. $4x^{\frac{5}{2}}y$

    c. $4xy$

    d. $32x^{\frac{7}{2}}y^{\frac{11}{2}}$

    e. x^3y^6

60. A ball is thrown from the top of a high hill, so that the height of the ball as a function of time is $h(t) = -16t^2 + 4t + 6$, in feet. What is the maximum height of the ball in feet?

    a. 6

    b. 6.25

    c. 6.5

    d. 6.75

    e. 6.8

Answer Explanations #1

1. D: This problem can be solved using basic arithmetic. Xavier starts with 20 apples, then gives his sister half, so 20 divided by 2.

$$\frac{20}{2} = 10$$

He then gives his neighbor 6, so 6 is subtracted from 10.

$$10 - 6 = 4$$

Lastly, he uses ¾ of his apples to make an apple pie, so to find remaining apples, the first step is to subtract ¾ from one and then multiply the difference by 4.

$$\left(1 - \frac{3}{4}\right) \times 4 = ?$$

$$\left(\frac{4}{4} - \frac{3}{4}\right) \times 4 = ?$$

$$\left(\frac{1}{4}\right) \times 4 = 1$$

2. C: The product of two irrational numbers can be rational or irrational. Sometimes, the irrational parts of the two numbers cancel each other out, leaving a rational number. For example, $\sqrt{2} * \sqrt{2} = 2$ because the roots cancel each other out. Technically, the product of two irrational numbers can be complex because complex numbers can have either the real or imaginary part (in this case, the imaginary part) equal zero and still be considered a complex number. However, Choice *D* is incorrect because the product of two irrational numbers is not an imaginary number so saying the product is complex *and* imaginary is incorrect.

3. B: The car is traveling at a speed of five meters per second. On the interval from one to three seconds, the position changes by fifteen meters. By making this change in position over time into a rate, the speed becomes ten meters in two seconds or five meters in one second.

4. C: The number negative four is classified as a real number because it exists and is not imaginary. It is rational because it does not have a decimal that never ends. It is an integer because it does not have a fractional component. The next classification would be whole numbers, for which negative four does not qualify because it is negative. Although -4 could technically be considered a complex number because complex numbers can have either the real or imaginary part equal zero and still be considered a complex number, Choice *D* is wrong because -4 is not considered an irrational number because it does not have a never-ending decimal component.

5. B: Since $850 is the price *after* a 20% discount, $850 represents 80% of the original price. To determine the original price, set up a proportion with the ratio of the sale price (850) to original price (unknown) equal to the ratio of sale percentage:

$$\frac{850}{x} = \frac{80}{100}$$

(where *x* represents the unknown original price)

To solve a proportion, cross-multiply the numerators and denominators and set the products equal to each other: (850)(100) = (80)(x). Multiplying each side results in the equation 85,000 = 80x.

To solve for *x*, divide both sides by 80: $\frac{85,000}{80} = \frac{80x}{80}$, resulting in x = 1062.5. Remember that *x* represents the original price. Subtracting the sale price from the original price ($1062.50 – $850) indicates that Frank saved $212.50.

6. E: This problem can be multiplied as 588 × 32, except at the end, the decimal point needs to be moved three places to the left. Performing the multiplication will give 18,816 and moving the decimal place over three places results in 18.816.

7. C: The first step in solving this problem is expressing the result in fraction form. Separate this problem first by solving the division operation of the last two fractions. When dividing one fraction by another, invert or flip the second fraction and then multiply the numerator and denominator.

$$\frac{7}{10} \times \frac{2}{1} = \frac{14}{10}$$

Next, multiply the first fraction with this value:

$$\frac{3}{5} \times \frac{14}{10} = \frac{42}{50}$$

Decimals are expressions of 1 or 100%, so multiply both the numerator and denominator by 2 to get the fraction as an expression of 100.

$$\frac{42}{50} \times \frac{2}{2} = \frac{84}{100}$$

In decimal form, this would be expressed as 0.84.

8. E: To find the average of a set of values, add the values together and then divide by the total number of values. In this case, include the unknown value of what Dwayne needs to score on his next test, in order to solve it.

$$\frac{78 + 92 + 83 + 97 + x}{5} = 90$$

Add the unknown value to the new average total, which is 5. Then multiply each side by 5 to simplify the equation, resulting in:

$$78 + 92 + 83 + 97 + x = 450$$

$$350 + x = 450$$

$$x = 100$$

Dwayne would need to get a perfect score of 100 in order to get an average of at least 90.

Test this answer by substituting back into the original formula.

$$\frac{78 + 92 + 83 + 97 + 100}{5} = 90$$

9. D: 1, 2, 3, 4, 6, 12. A given number divides evenly by each of its factors to produce an integer (no decimals). The number 5, 7, 8, 9, 10, 11 (and their opposites) do not divide evenly into 12. Therefore, these numbers are not factors.

10. E: The augmented matrix that represents the system of equations has dimensions 4 x 3 because there are three equations with three unknowns. The coefficients of the variables make up the first three columns, and the last column is made up of the numbers to the right of the equals sign. This system can be solved by reducing the matrix to row-echelon form, where the last column gives the solution for the unknown variables.

11. B: There are two zeros for the function: $x = 0, -2$. The zeros can be found several ways, but this particular equation can be factored into $f(x) = x(x^2 + 4x + 4) = x(x + 2)(x + 2)$. By setting each factor equal to zero and solving for x, there are two solutions. On a graph these zeros can be seen where the line crosses the x-axis.

12. D: This problem involves a composition function, where one function is plugged into the other function. In this case, the $f(x)$ function is plugged into the $g(x)$ function for each x-value. The composition equation becomes $g(f(x)) = 2^3 - 3(2^2) - 2(2) + 6$. Simplifying the equation gives the answer $g(f(x)) = 8 - 3(4) - 2(2) + 6 = 8 - 12 - 4 + 6 = -2$.

13. D: This system of equations involves one quadratic function and one linear function, as seen from the degree of each equation. One way to solve this is through substitution. Solving for y in the second equation yields $y = x + 2$. Plugging this equation in for the y of the quadratic equation yields $x^2 - 2x + x + 2 = 8$. Simplifying the equation, it becomes $x^2 - x + 2 = 8$. Setting this equal to zero and factoring, it becomes $x^2 - x - 6 = 0 = (x - 3)(x + 2)$. Solving these two factors for x gives the zeros $x = 3, -2$. To find the y-value for the point, each number can be plugged in to either original equation. Solving each one for y yields the points $(3, 5)$ and $(-2, 0)$.

14. D: The expression is simplified by collecting like terms. Terms with the same variable and exponent are like terms, and their coefficients can be added.

15. A: Finding the product means distributing one polynomial onto the other. Each term in the first must be multiplied by each term in the second. Then, like terms can be collected. Multiplying the factors yields the expression $20x^3 + 4x^2 + 24x - 40x^2 - 8x - 48$. Collecting like terms means adding the x^2 terms and adding the x terms. The final answer after simplifying the expression is $20x^3 - 36x^2 + 16x - 48$.

16. E: Finding the zeros for a function by factoring is done by setting the equation equal to zero, then completely factoring. Since there was a common x for each term in the provided equation, that is factored out first. Then the quadratic that is left can be factored into two binomials: $(x + 1)(x - 4)$. Setting each factor equation equal to zero and solving for x yields three zeros.

17. A: 847.90. The hundredth place value is located two digits to the right of the decimal point (the digit 9). The digit to the right of the place value is examined to decide whether to round up or keep the digit. In this case, the digit 6 is 5 or greater so the hundredth place is rounded up. When rounding up, if the digit to be increased is a 9, the digit to its left is increased by one and the digit in the desired place value is made a zero. Therefore, the number is rounded to 847.90.

18. E: The equation can be solved by factoring the numerator into $(x + 6)(x - 5)$. Since that same factor exists on top and bottom, that factor $(x - 5)$ cancels. This leaves the equation $x + 6 = 11$. Solving the equation gives the answer $x = 5$. When this value is plugged into the equation, it yields a zero in the denominator of the fraction. Since this is undefined, there is no solution.

19. D: This problem can be solved by using unit conversion. The initial units are miles per minute. The final units need to be feet per second. Converting miles to feet uses the equivalence statement 1 mile = 5,280 feet. Converting minutes to seconds uses the equivalence statement 1 minute = 60 seconds. Setting up the ratios to convert the units is shown in the following equation $\frac{72\ miles}{90\ minutes} * \frac{1\ minute}{60\ seconds} *$ $\frac{5280\ feet}{1\ mile} = 70.4$ feet per second. The initial units cancel out, and the new units are left.

20. B: The formula can be manipulated by dividing both the length, *l*, and the width, *w*, on both sides. The length and width will cancel on the right, leaving height by itself.

21. B: The domain is all possible input values, or x-values. For this equation, the domain is every number greater than or equal to zero. There are no negative numbers in the domain because taking the square root of a negative number results in an imaginary number.

22. D: This problem can be solved by setting up a proportion involving the given information and the unknown value. The proportion is $\frac{21\ pages}{4\ night} = \frac{140\ pages}{x\ nigh}$. Solving the proportion by cross-multiplying, the equation becomes $21x = 4 * 140$, where $x = 26.67$. Since it is not an exact number of nights, the answer is rounded up to 27 nights. Twenty-six nights would not give Sarah enough time.

23. D: The slope from this equation is 50, and it is interpreted as the cost per gigabyte used. Since the g-value represents number of gigabytes and the equation is set equal to the cost in dollars, the slope relates these two values. For every gigabyte used on the phone, the bill goes up 50 dollars.

24. C: Graphing the function $y = \cos(x)$ shows that the curve starts at $(0, 1)$, has an amplitude of 2, and a period of 2π. This same curve can be constructed using the sine graph, by shifting the graph to the left $\frac{\pi}{2}$ units. This equation is in the form $y = \sin(x + \frac{\pi}{2})$.

25. B: Start by squaring both sides to get $1 + x = 16$. Then subtract 1 from both sides to get $x = 15$.

26. B: This inverse of a function is found by switching the x and y in the equation and solving for y. In the given equation, solving for y is done by adding 5 to both sides, then dividing both sides by 3. This answer can be checked on the graph by verifying the lines are reflected over $y = x$.

27. B: The zeros of this function can be found by using the quadratic formula, $x = \frac{-b \pm \sqrt{b^2 - 4ac}}{2a}$. Identifying *a*, *b*, and *c* can also be done from the equation because it is in standard form. The formula becomes $x = \frac{0 \pm \sqrt{0^2 - 4(1)(4)}}{2(1)} = \frac{\sqrt{-16}}{2}$. Since there is a negative underneath the radical, the answer is a complex number.

28. E: Setting up a proportion is the easiest way to represent this situation. The proportion becomes $\frac{20}{x} = \frac{40}{100}$, where cross-multiplication can be used to solve for x. The answer can also be found by observing the two fractions as equivalent, knowing that twenty is half of forty, and fifty is half of one-hundred.

29. C: Division with scientific notation can be solved by grouping the first terms together and grouping the tens together. The first terms can be divided, and the tens terms can be simplified using the rules for exponents. The initial expression becomes $0.4 * 10^4$. This is not in scientific notation because the first number is not between 1 and 10. Shifting the decimal and subtracting one from the exponent yields $4.0 * 10^3$.

30. A: The relative error can be found by finding the absolute error and making it a percent of the true value. The absolute error is $36 - 35.75 = 0.25$. This error is then divided by 36—the true value—to find 0.7%.

31. B: The y-intercept of an equation is found where the x-value is zero. Plugging zero into the equation for x allows the first two terms to cancel out, leaving -4.

32. A: The equation is *even* because $f(-x) = f(x)$. Plugging in a negative value will result in the same answer as when plugging in the positive of that same value. The function $f(-2) = \frac{1}{2}(-2)^4 + 2(-2)^2 - 6 = 8 + 8 - 6 = 10$ yields the same value as $f(2) = \frac{1}{2}(2)^4 + 2(2)^2 - 6 = 8 + 8 - 6 = 10$.

33. C: The equation $x = 3$ is not a function because it does not pass the vertical line test. This test is made from the definition of a function, where each x-value must be mapped to one, and only one, y-value. This equation is a vertical line, so the x-value of 3 is mapped with an infinite number of y-values.

34. A: The function can be factored to identify the zeros. First, the term $3x$ is factored out to the front because each term contains $3x$. Then, the quadratic is factored into $(x + 3)(x - 2)$.

35. C: A die has an equal chance for each outcome. Since it has six sides, each outcome has a probability of $\frac{1}{6}$. The chance of a 1 or a 2 is therefore $\frac{1}{6} + \frac{1}{6} = \frac{1}{3}$.

36. A: The slope is given by $m = \frac{y_2 - y_1}{x_2 - x_1} = \frac{0 - 4}{0 - (-3)} = -\frac{4}{3}$.

37. B: The table shows values that are increasing exponentially. The differences between the inputs are the same, while the differences in the outputs are changing by a factor of 2. The values in the table can be modeled by the equation $f(x) = 2^x$.

38. C: The formula for continually compounded interest is $A = Pe^{rt}$. Plugging in the given values to find the total amount in the account yields the equation $A = 2000e^{0.05 * 8} = 2983.65$.

39. E: The sample space is made up of $8 + 7 + 6 + 5 = 26$ balls. The probability of pulling each individual ball is $^1/_{26}$. Since there are 7 yellow balls, the probability of pulling a yellow ball is $^7/_{26}$.

40. B: For the first card drawn, the probability of a King being pulled is $^4/_{52}$. Since this card isn't replaced, if a King is drawn first, the probability of a King being drawn second is $^3/_{51}$. The probability of a King being drawn in both the first and second draw is the product of the two probabilities: $^4/_{52}$ x $^3/_{51}= {}^{12}/_{2652}$ which, divided by 12, equals $^1/_{221}$.

41. E: The addition rule is necessary to determine the probability because a 6 can be rolled on either roll of the die. The rule used is $P(A \text{ or } B) = P(A) + P(B) - P(A \text{ and } B)$. The probability of a 6 being individually rolled is $^1/_6$ and the probability of a 6 being rolled twice is $^1/_6 \cdot {}^1/_6 = {}^1/_{36}$. Therefore, the probability that a 6 is rolled at least once is $1/6 + 1/6 - 1/36 = 11/36$

42. A: $(A \cap B)$ is equal to the intersection of the two sets A and B, which is $\{1, 2, 3, 4, 5\}$. $A - (A \cap B)$ is equal to the elements of A that are *not* included in the set $(A \cap B)$. Therefore, $A - (A \cap B) = \{6, 7, 8, 9, 10\}$.

43. B: An equilateral triangle has three sides of equal length, so if the total perimeter is 18 feet, each side must be 6 feet long. A square with sides of 6 feet will have an area of $6^2 = 36$ square feet.

44. A: If each man gains 10 pounds, every original data point will increase by 10 pounds. Therefore, the man with the original median will still have the median value, but that value will increase by 10. The smallest value and largest value will also increase by 10 and, therefore, the difference between the two won't change. The range does not change in value and, thus, remains the same.

45. C: Because the triangles are similar, the lengths of the corresponding sides are proportional. Therefore:

$$\frac{30 + x}{30} = \frac{22}{14} = \frac{y + 15}{y}$$

This results in the equation $14(30 + x) = 22 \cdot 30$ which, when solved, gives $x = 17.1$. The proportion also results in the equation $14(y + 15) = 22y$ which, when solved, gives $y = 26.3$.

46. A: Robert accomplished his task on Tuesday in ¾ the time compared to Monday. He must have worked 4/3 as fast.

47. B: To simplify this inequality, subtract 3 from both sides to get $-\frac{1}{2}x \geq -1$. Then, multiply both sides by -2 (remembering this flips the direction of the inequality) to get $x \leq 2$.

48. E: There are two ways to approach this problem. Each value can be substituted into each equation. Choice *A* can be eliminated, since $4^2 + 16 = 32$. Choice *B* can be eliminated, since $4^2 + 4 \cdot 4 - 4 = 28$. Choice *C* can be eliminated, since $4^2 - 2 \cdot 4 - 2 = 6$. But, plugging in either value into $x^2 - 16$ gives $(\pm 4)^2 - 16 = 16 - 16 = 0$.

49. B: The technique of completing the square must be used to change $4x^2 + 4y^2 - 16x - 24y + 51 = 0$ into the standard equation of a circle. First, the constant must be moved to the right-hand side of the equals sign, and each term must be divided by the coefficient of the x^2 term (which is 4). The x and y

terms must be grouped together to obtain $x^2 - 4x + y^2 - 6y = -\frac{51}{4}$. Next, the process of completing the square must be completed for each variable. This gives $(x^2 - 4x + 4) + (y^2 - 6y + 9) = -\frac{51}{4} + 4 + 9$. The equation can be written as $(x - 2)^2 + (y - 3)^2 = \frac{1}{4}$. Therefore, the center of the circle is (2, 3) and the radius is $\sqrt{1/4} = 1/2$.

50. D: When an ordered pair is reflected over an axis, the sign of one of the coordinates must change. When it's reflected over the x-axis, the sign of the y-coordinate must change. The x-value remains the same. Therefore, the new ordered pair is $(-3, 4)$.

51. A: Because the volume of the given sphere is 288π cubic meters, this means $\frac{4}{3}\pi r^3 = 288\pi$. This equation is solved for r to obtain a radius of 6 meters. The formula for the surface area of a sphere is $4\pi r^2$, so if $r = 6$ in this formula, the surface area is 144π square meters.

52. E: SOHCAHTOA is used to find the missing side length. Because the angle and adjacent side are known, $\tan 60 = \frac{x}{13}$. Making sure to evaluate tangent with an argument in degrees, this equation gives $x = 13\tan 60 = 13 \cdot 1.73 = 22.49$.

53. A: Let the unknown score be x. The average will be $\frac{5 \cdot 50 + 4 \cdot 70 + x}{10} = \frac{530 + x}{10} = 55$. Multiply both sides by 10 to get $530 + x = 550$, or $x = 20$.

54. A: First, the sample mean must be calculated. $\bar{x} = \frac{1}{4}(1 + 3 + 5 + 7) = 4$. The standard deviation of the data set is $\sigma = \sqrt{\frac{\Sigma(x - \bar{x})^2}{n - 1}}$, and $n = 4$ represents the number of data points. Therefore:

$$\sigma = \sqrt{\frac{1}{3}[(1 - 4)^2 + (3 - 4)^2 + (5 - 4)^2 + (7 - 4)^2]} = \sqrt{\frac{1}{3}(9 + 1 + 1 + 9)} = 2.58$$

55. D: For manufacturing costs, there is a linear relationship between the cost to the company and the number produced, with a y-intercept given by the base cost of acquiring the means of production, and a slope given by the cost to produce one unit. In this case, that base cost is \$50,000, while the cost per unit is \$40. So, $y = 40x + 50,000$.

56. C: The expected value is equal to the total sum of each product of individual score and probability. There are 36 possible rolls. The probability of rolling a 2 is $\frac{1}{36}$. The probability of rolling a 3 is $\frac{2}{36}$. The probability of rolling a 4 is $\frac{3}{36}$. The probability of rolling a 5 is $\frac{4}{36}$. The probability of rolling a 6 is $\frac{5}{36}$. The probability of rolling a 7 is $\frac{6}{36}$. The probability of rolling an 8 is $\frac{5}{36}$. The probability of rolling a 9 is $\frac{4}{36}$. The probability of rolling a 10 is $\frac{3}{36}$. The probability of rolling an 11 is $\frac{2}{36}$. Finally, the probability of rolling a 12 is $\frac{1}{36}$.

Each possible outcome is multiplied by the probability of it occurring. Like this:

$$2 \times \frac{1}{36} = a$$

$$3 \times \frac{2}{36} = b$$

$$4 \times \frac{3}{36} = c$$

And so forth.

Then all of those results are added together:

$$a + b + c \ldots = expected\ value$$

In this case, it equals 7.

57. B: From the slope-intercept form, $y = mx + b$, it is known that b is the y-intercept, which is 1. Compute the slope as $\frac{2-1}{1-0} = 1$, so the equation should be $y = x + 1$.

58. A: This has the form $t^2 - y^2$, with $t = x^2$ and $y = 4$. It's also known that $t^2 - y^2 = (t + y)(t - y)$, and substituting the values for t and y into the right-hand side gives $(x^2 - 4)(x^2 + 4)$.

59. A: Simplify this to $(4x^2y^4)^{\frac{3}{2}} = 4^{\frac{3}{2}}(x^2)^{\frac{3}{2}}(y^4)^{\frac{3}{2}}$. Now, $4^{\frac{3}{2}} = (\sqrt{4})^3 = 2^3 = 8$. For the other, recall that the exponents must be multiplied, so this yields $8x^{2 \cdot \frac{3}{2}}y^{4 \cdot \frac{3}{2}} = 8x^3y^6$.

60. B: The independent variable's coordinate at the vertex of a parabola (which is the highest point, when the coefficient of the squared independent variable is negative) is given by $x = -\frac{b}{2a}$. Substitute and solve for x to get $x = -\frac{4}{2(-16)} = \frac{1}{8}$. Using this value of x, the maximum height of the ball (y), can be calculated. Substituting x into the equation yields $h(t) = -16\frac{1}{8}^2 + 4\frac{1}{8} + 6 = 6.25$.

ACT Math Practice Test #2

1. What is the solution to 4 x 7 + (25 − 21)² ÷ 2?

    a. 512
    b. 36
    c. 60.5
    d. 22
    e. 16

2. Johnny earns $2334.50 from his job each month. He pays $1437 for monthly expenses. Johnny is planning a vacation in 3 months' time that he estimates will cost $1750 total. How much will Johnny have left over from three months' of saving once he pays for his vacation?

    a. $948.50
    b. $584.50
    c. $852.50
    d. $942.50
    e. $848.50

3. What is the solution to $(2 \times 20) \div (7 + 1) + (6 \times 0.01) + (4 \times 0.001)$?

    a. 5.064
    b. 5.64
    c. 5.0064
    d. 48.064
    e. 56.4

4. A piggy bank contains 12 dollars' worth of nickels. A nickel weighs 5 grams, and the empty piggy bank weighs 1050 grams. What is the total weight of the full piggy bank?

    a. 1,110 grams
    b. 1,200 grams
    c. 2,110 grams
    d. 2,200 grams
    e. 2,250 grams

5. Last year, the New York City area received approximately $27 \, ^3/_4$ inches of snow. The Denver area received approximately 3 times as much snow as New York City. How much snow fell in Denver?

    a. 60 inches
    b. $27 \, ^1/_4$ inches
    c. $9 \, ^1/_4$ inches
    d. $83 \, ^1/_4$ inches
    e. $29 \, ^1/_4$ inches

6. Which of the following correctly arranges the numbers from least to greatest value?

$0.85, \frac{4}{5}, \frac{2}{3}, \frac{91}{100}$

    a. $0.85, \frac{4}{5}, \frac{2}{3}, \frac{91}{100}$

    b. $\frac{4}{5}, 0.85, \frac{91}{100}, \frac{2}{3}$

    c. $\frac{2}{3}, \frac{4}{5}, 0.85, \frac{91}{100}$

    d. $0.85, \frac{91}{100}, \frac{4}{5}, \frac{2}{3}$

    e. $\frac{4}{5}, \frac{2}{3}, 0.85, \frac{91}{100}$

7. Four people split a bill. The first person pays for $\frac{1}{5}$, the second person pays for $\frac{1}{4}$, and the third person pays for $\frac{1}{3}$. What fraction of the bill does the fourth person pay?

    a. $\frac{1}{12}$

    b. $\frac{47}{60}$

    c. $\frac{1}{4}$

    d. $\frac{4}{15}$

    e. $\frac{13}{60}$

8. In a school with 300 students, there are 10 students with red hair, 50 students with black hair, 180 students with brown hair, and 60 students with blonde hair. What is the ratio of blonde hair to brown hair?

    a. 3:1
    b. 2:1
    c. 1:3
    d. 1:2
    e. 1:5

9. What is the value of b in this equation? $5b - 4 = 2b + 17$

    a. 13
    b. 24
    c. 7
    d. 21
    e. 9

10. If $g(x) = x^3 - 3x^2 - 2x + 6$ and $f(x) = 2$, then what is $g(f(x))$?

    a. -26

    b. 6

    c. $2x^3 - 6x^2 - 4x + 12$

    d. $-3x - 2$

    e. -2

11. What is the value of the sum of $\frac{1}{3}$ and $\frac{2}{5}$?

    a. $\frac{3}{8}$

    b. $\frac{11}{15}$

    c. $\frac{11}{30}$

    d. $\frac{4}{5}$

    e. $\frac{2}{8}$

12. What is the solution for the equation $\tan(x) + 1 = 0$, where $0 \leq x < 2\pi$?

    a. $x = \frac{3\pi}{4}, \frac{5\pi}{4}$

    b. $x = \frac{3\pi}{4}, \frac{\pi}{4}$

    c. $x = \frac{5\pi}{4}, \frac{7\pi}{4}$

    d. $x = \frac{3\pi}{4}, \frac{7\pi}{4}$

    e. $x = \frac{5\pi}{4}, \frac{\pi}{4}$

13. What are the first four terms of the series $\left\{ \frac{(-1)^{n+1}}{n^2+5} \right\}_{n=0}^{\infty}$?

    a. $\frac{1}{6}, \frac{1}{9}, \frac{1}{14}, \frac{1}{19}$

    b. $\frac{1}{6}, \frac{-1}{9}, \frac{1}{14}, \frac{-1}{19}$

    c. $\frac{-1}{5}, \frac{1}{6}, \frac{-1}{9}, \frac{1}{14}$

    d. $\frac{1}{5}, \frac{1}{6}, \frac{1}{9}, \frac{1}{14}$

    e. $\frac{1}{5}, -\frac{1}{6}, -\frac{1}{9}, \frac{1}{14}$

14. A particle moves along the x-axis so that at any time $t \geq 0$, its velocity is given by $v(t) = \frac{6}{t+3}$. What is the acceleration of the particle at time $t = 5$?

    a. $-\frac{2}{3}$

    b. $-\frac{3}{16}$

    c. $\frac{3}{4}$

    d. $\frac{2}{3}$

    e. $-\frac{3}{32}$

15. Divide and reduce $4/13 \div 27/169$.

    a. 52/27

    b. 51/27

    c. 52/29

    d. 51/29

    e. 7/13

16. Is the series $\sum_{k=0}^{\infty}(-1)^{k}\left(\frac{2}{3}\right)^{k}$ convergent or divergent? If convergent, find its sum.

    a. Divergent

    b. Convergent, $\frac{3}{5}$

    c. Convergent, $\frac{5}{3}$

    d. Convergent, $\frac{2}{3}$

    e. Convergent, $\frac{3}{2}$

17. Katie works at a clothing company and sold 192 shirts over the weekend. $\frac{1}{3}$ of the shirts that were sold were patterned, and the rest were solid. Which mathematical expression would calculate the number of solid shirts Katie sold over the weekend?

    a. $192 \times \frac{1}{3}$

    b. $192 \div \frac{1}{3}$

    c. $192 \times \left(1 - \frac{1}{3}\right)$

    d. $192 \div 3$

    e. $192 \times \left(-\frac{1}{3}\right)$

18. Which of the following is the result of simplifying the expression:
$$\frac{4a^{-1}b^3}{a^4b^{-2}} * \frac{3a}{b}$$

    a. $12a^3b^5$

    b. $12\frac{b^2}{a^2}$

    c. $\frac{12}{a^4}$

    d. $7\frac{b^4}{a}$

    e. $12\frac{b^4}{a^4}$

19. What is the simplified quotient of the following expression?
$$\frac{5x^3}{3x^2y} \div \frac{25}{3y^9}$$

    a. $\frac{125}{9y^{10}}$

    b. $\frac{x}{5y^8}$

    c. $\frac{5}{xy^8}$

    d. $\frac{xy^8}{5}$

    e. $\frac{xy^2}{5x}$

20. What is the volume of a cube with the side equal to 3 inches?
    a. 6 in³
    b. 27 in³
    c. 9 in³
    d. 3 in³
    e. 81 in³

21. What is the volume of a cube with the side equal to 5 centimeters?
    a. 10 cm³
    b. 15 cm³
    c. 50 cm³
    d. 25 cm³
    e. 125 cm³

22. What is the volume of a rectangular prism with a height of 2 inches, a width of 4 inches, and a depth of 6 inches?
    a. 12 in³
    b. 24 in³
    c. 48 in³
    d. 60 in³
    e. 96 in³

23. What is the volume of a rectangular prism with the height of 3 centimeters, a width of 5 centimeters, and a depth of 11 centimeters?
    a. 19 cm³
    b. 165 cm³
    c. 225 cm³
    d. 150 cm³
    e. 88 cm³

24. What is the volume of a pyramid, with the area of the base measuring 12 inches², and the height measuring 15 inches?
    a. 180 in³
    b. 90 in³
    c. 30 in³
    d. 60 in³
    e. 45 in³

25. A pizzeria owner regularly creates jumbo pizzas, each with a radius of 9 inches. She is mathematically inclined, and wants to know the area of the pizza to purchase the correct boxes and know how much she is feeding her customers. What is the area of the circle, in terms of π, with a radius of 9 inches?
    a. 3 π in²
    b. 18 π in²
    c. 90 π in²
    d. 9 π in²
    e. 81 π in²

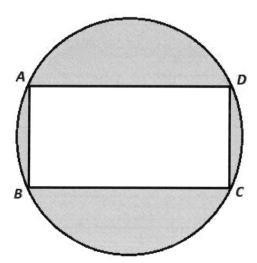

26. Rectangle *ABCD* is inscribed in the circle above. The length of side *AB* is 9 inches and the length of side *BC* is 12 inches. What is the area of the shaded region?
    a. 64.4 sq. in.
    b. 68.6 sq. in.
    c. 62.8 sq. in.
    d. 61.3 sq. in.
    e. 64.6 sq. in.

27. Using the following diagram, calculate the total circumference, rounding to the nearest decimal place:

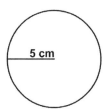

5 cm

   a. 25.0 cm
   b. 15.7 cm
   c. 78.5 cm
   d. 50.0 cm
   e. 31.4 cm

28. An angle measures 54 degrees. In order to correctly determine the measure of its complementary angle, what concept is necessary?
   a. Two complementary angles sum up to 180 degrees.
   b. Complementary angles are always acute.
   c. Two complementary angles sum up to 90 degrees.
   d. Complementary angles sum up to 360 degrees.
   e. At least one of the angles of two complementary angles must be obtuse.

29. A triangular pyramid is composed of all equilateral triangles with sides measuring 8 cm. What is the surface area of the pyramid?
   a. 128 sq. cm.
   b. 110.7 sq. cm.
   c. 114.6 sq. cm.
   d. 117.8 sq. cm.
   e. 27.68 sq. cm.

30. What's the midpoint of a line segment with endpoints $(-1, 2)$ and $(3, -6)$?
   a. $(1, 2)$
   b. $(1, 0)$
   c. $(-1, 2)$
   d. $(1, -2)$
   e. $(1, 4)$

31. Given the following triangle, what's the length of the missing side? Round the answer to the nearest tenth.

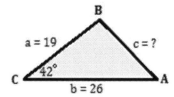

a. 17.0
b. 17.4
c. 18.0
d. 18.4
e. 17.2

32. What are the coordinates of the focus of the parabola $y = -9x^2$?
   a. $(-3, 0)$
   b. $\left(-\frac{1}{36}, 0\right)$
   c. $(0, -3)$
   d. $\left(-3, -\frac{1}{36}\right)$
   e. $\left(0, -\frac{1}{36}\right)$

33. According to building code regulations, the roof of a house has to be set at a minimum angle of 39° up to a maximum angle of 48° to ensure snow and rain will properly slide off it. What is the maximum incline in terms of radians?
   a. $\frac{\pi}{4}$
   b. $\frac{\pi}{15}$
   c. $\frac{4\pi}{15}$
   d. $\frac{3\pi}{4}$
   e. $\frac{48\pi}{90}$

34. An arc is intercepted by a central angle of 240°. What is the number of radians of that angle?
   a. $\frac{3\pi}{4}$
   b. $\frac{4\pi}{3}$
   c. $\frac{\pi}{4}$
   d. $\frac{\pi}{3}$
   e. $\frac{3}{4\pi}$

35. Let p = "Alex is an engineering major," q = "Alex is not an English major," r = "Alex's sister is a history major," s = "Alex's sister has been to Germany," and t = "Alex's sister has been to Austria." Which of the following answers represents the statement "Alex is an engineering and English major, but his sister is a history major who hasn't been to either Germany or Austria."?

    a. $p \wedge \sim q \wedge (r \vee (\sim s \vee \sim t))$

    b. $p \wedge q \wedge r \vee (\sim s \wedge \sim t)$

    c. $p \wedge \sim q \wedge r \wedge (\sim s \vee \sim t)$

    d. $p \wedge q \wedge (r \vee (\sim s \wedge \sim t))$

    e. $p \wedge \sim q \wedge r \vee (\sim s \wedge \sim t)$

36. Find the determinant of the matrix $\begin{bmatrix} -4 & 2 \\ 3 & -1 \end{bmatrix}$.

    a. -10

    b. -2

    c. 0

    d. 2

    e. -1

37. In a statistical experiment, 29 college students are given an exam during week 11 of the semester, and 30 college students are given an exam during week 12 of the semester. Both groups are being tested to determine which exam week might result in a higher grade. What's the degree of freedom in this experiment?

    a. 29

    b. 30

    c. 59

    d. 28

    e. 23

38. What is the probability of randomly picking the winner and runner-up from a race of 4 horses and distinguishing which is the winner?

    a. $\frac{1}{4}$

    b. $\frac{1}{2}$

    c. $\frac{1}{16}$

    d. $\frac{1}{12}$

    e. $\frac{1}{64}$

39. What is the interquartile range (IQR) of the following data set? 1, 4, 6, 6, 9, 10, 12, 17, 18

    a. 10

    b. 9

    c. 14.5

    d. 5

    e. 9.5

$$G = .035O + .26$$

40. The linear regression model above is based on an analysis of the price of a gallon of gas (G) at 15 gas stations compared to the price of a barrel of oil (O) at the time. Based on this model, which of the following statements are true?

      I. There is a negative correlation between G and O.

      II. When oil is $55 per barrel then gas is approximately $2.19 per gallon.

      III. The slope of the line indicates that as O increases by 1, G increases by .035.

      IV. If the price of oil increases by $8 per barrel then the price of gas will increase by approximately $0.18 per gallon.

  a. I and II

  b. II only

  c. II and III

  d. I and III

  e. II, III, and IV

41. Given the value of a given stock at monthly intervals, which graph should be used to best represent the trend of the stock?

  a. Box plot

  b. Line plot

  c. Line graph

  d. Circle graph

  e. Dot plot

42. Alan currently weighs 200 pounds, but he wants to lose weight to get down to 175 pounds. What is this difference in kilograms? (1 pound is approximately equal to 0.45 kilograms.)

  a. 9 kg

  b. 11.25 kg

  c. 78.75 kg

  d. 90 kg

  e. 25 kg

43. What is $\frac{420}{98}$ rounded to the nearest integer?

  a. 3

  b. 4

  c. 5

  d. 6

  e. 7

44. The following graph compares the various test scores of the top three students in each of these teacher's classes. Based on the graph, which teacher's students had the smallest range of test scores?

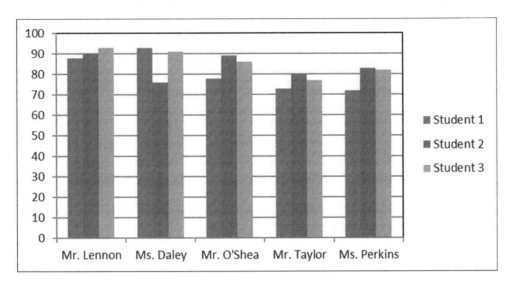

a. Mr. Lennon
b. Mr. O'Shea
c. Mr. Taylor
d. Ms. Daley
e. Ms. Perkins

45. The width of a rectangular house is 22 feet. What is the perimeter of this house if it has the same area as a house that is 33 feet wide and 50 feet long?
a. 184 feet
b. 200 feet
c. 192 feet
d. 206 feet
e. 194 feet

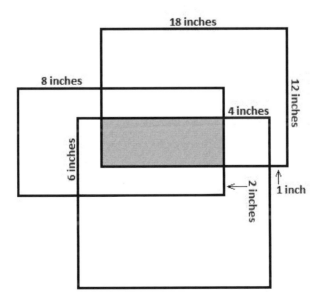

46. In the figure above, what is the area of the shaded region?
   a. 48 sq. inches
   b. 52 sq. inches
   c. 44 sq. inches
   d. 56 sq. inches
   e. 46 sq. inches

47. If $3x = 6y = -2z = 24$, then what does $4xy + z$ equal?
   a. 116
   b. 130
   c. 84
   d. 108
   e. 98

48. If $n = 2^2$, and $m = n^2$, then m^n equals?
   a. 2^{12}
   b. 2^{10}
   c. 2^{18}
   d. 2^{16}
   e. 2^{20}

49. $\frac{3}{25} =$
   a. 0.15
   b. 0.1
   c. 0.9
   d. 0.12
   e. 0.19

50. Which of the following is largest?
   a. 0.45
   b. 0.096
   c. 0.3
   d. 0.313
   e. 0.299

51. Which of the following is NOT a way to write 40 percent of N?
   a. $(0.4)N$
   b. $\frac{2}{5}N$
   c. $N \times 0.4$
   d. $\frac{4N}{10}$
   e. $40N$

52. Which is closest to 17.8×9.9?
   a. 140
   b. 180
   c. 200
   d. 350
   e. 400

53. A student gets an 85% on a test with 20 questions. How many answers did the student solve correctly?

    a. 15

    b. 16

    c. 17

    d. 18

    e. 19

54. 6 is 30% of what number?

    a. 18

    b. 20

    c. 24

    d. 25

    e. 26

55. $3\frac{2}{3} - 1\frac{4}{5} =$

    a. $1\frac{13}{15}$

    b. $\frac{14}{15}$

    c. $2\frac{2}{3}$

    d. $\frac{4}{5}$

    e. $\frac{4}{15}$

56. $4\frac{1}{3} + 3\frac{3}{4} =$

    a. $6\frac{5}{12}$

    b. $8\frac{1}{12}$

    c. $8\frac{2}{3}$

    d. $7\frac{7}{12}$

    e. $7\frac{4}{12}$

57. Five of six numbers have a sum of 25. The average of all six numbers is 6. What is the sixth number?

    a. 8

    b. 10

    c. 13

    d. 12

    e. 11

58. $52.3 \times 10^{-3} =$

    a. 0.00523

    b. 0.0523

    c. 0.523

    d. 523

    e. 5.23

59. If $\dfrac{5}{2} \div \dfrac{1}{3} = n$, then n is between:

    a. 5 and 7

    b. 7 and 9

    c. 9 and 11

    d. 3 and 5

    e. 11 and 13

60. Shawna buys $2\dfrac{1}{2}$ gallons of paint. If she uses $\dfrac{1}{3}$ of it on the first day, how much does she have left?

    a. $1\dfrac{5}{6}$ gallons

    b. $1\dfrac{1}{2}$ gallons

    c. $1\dfrac{1}{3}$ gallons

    d. 2 gallons

    e. $1\dfrac{2}{3}$ gallons

Answer Explanations #2

1. B: To solve this correctly, keep in mind the order of operations with the mnemonic PEMDAS (Please Excuse My Dear Aunt Sally). This stands for Parentheses, Exponents, Multiplication, Division, Addition, Subtraction. Taking it step by step, solve the parentheses first:

$4 \times 7 + (4)^2 \div 2$

Then, apply the exponent:

$4 \times 7 + 16 \div 2$

Multiplication and division are both performed next:

$28 + 8 = 36$

2. D: First, subtract $1437 from $2334.50 to find Johnny's monthly savings; this equals $897.50. Then, multiply this amount by 3 to find out how much he will have (in three months) before he pays for his vacation: this equals $2692.50. Finally, subtract the cost of the vacation ($1750) from this amount to find how much Johnny will have left: $942.50.

3. A: Operations within the parentheses must be completed first. Then, division is completed. Finally, addition is the last operation to complete. When adding decimals, digits within each place value are added together. Therefore, the expression is evaluated as $(2 \times 20) \div (7 + 1) + (6 \times 0.01) + (4 \times 0.001) = 40 \div 8 + 0.06 + 0.004 = 5 + 0.06 + 0.004 = 5.064$.

4. E: A dollar contains 20 nickels. Therefore, if there are 12 dollars' worth of nickels, there are $12 \times 20 = 240$ nickels. Each nickel weighs 5 grams. Therefore, the weight of the nickels is $240 \times 5 = 1,200$ grams. Adding in the weight of the empty piggy bank, the filled bank weighs 2,250 grams.

5. D: 3 must be multiplied times $27\,^3/_4$. In order to easily do this, the mixed number should be converted into an improper fraction. $27\,^3/_4 = \frac{27 * 4 + 3}{4} = {}^{111}/_4$. Therefore, Denver had approximately $3 \times {}^{111}/_4 = {}^{333}/_4$ inches of snow. The improper fraction can be converted back into a mixed number through division. ${}^{333}/_4 = 83\,^1/_4$ inches.

6. C: The first step is to depict each number using decimals. $\frac{91}{100} = 0.91$

Multiplying both the numerator and denominator of $\frac{4}{5}$ by 20 makes it $\frac{80}{100}$ or 0.80; the closest approximation of $\frac{2}{3}$ would be $\frac{66}{100}$ or 0.66 recurring. Rearrange each expression in ascending order, as found in answer C.

7. E: To find the fraction of the bill that the first three people pay, the fractions need to be added, which means finding common denominator. The common denominator will be 60. $\frac{1}{5} + \frac{1}{4} + \frac{1}{3} = \frac{12}{60} + \frac{15}{60} + \frac{20}{60} = \frac{47}{60}$. The remainder of the bill is $1 - \frac{47}{60} = \frac{60}{60} - \frac{47}{60} = \frac{13}{60}$.

8. C: There are 60 students with blonde hair and 180 students with brown hair. The ratio would be set up as blonde:brown, so 60:180. When reduced this is 1:3. This means that for every 1 student with blonde hair, there are 3 with brown hair.

9. C: To solve for the value of b, both sides of the equation need to be equalized.

Start by cancelling out the lower value of -4 by adding 4 to both sides:

$5b - 4 = 2b + 17$
$5b - 4 + 4 = 2b + 17 + 4$
$5b = 2b + 21$

The variable b is the same on each side, so subtract the lower 2b from each side:

$5b = 2b + 21$
$5b - 2b = 2b + 21 - 2b$
$3b = 21$

Then divide both sides by 3 to get the value of b:

$$3b = 21$$

$$\frac{3b}{3} = \frac{21}{3}$$

$$b = 7$$

10. E: This problem involves a composition function, where one function is plugged into the other function. In this case, the $f(x)$ function is plugged into the $g(x)$ function for each x-value. The composition equation becomes $g(f(x)) = 2^3 - 3(2^2) - 2(2) + 6$. Simplifying the equation gives the answer $g(f(x)) = 8 - 3(4) - 2(2) + 6 = 8 - 12 - 4 + 6 = -2$.

11. B: $\frac{11}{15}$. Fractions must have like denominators to be added. The least common multiple of the denominators 3 and 5 is found. The LCM is 15, so both fractions should be changed to equivalent fractions with a denominator of 15. To determine the numerator of the new fraction, the old numerator is multiplied by the same number by which the old denominator is multiplied to obtain the new denominator. For the fraction $\frac{1}{3}$, 3 multiplied by 5 will produce 15. Therefore, the numerator is multiplied by 5 to produce the new numerator $\left(\frac{1 \times 5}{3 \times 5} = \frac{5}{15}\right)$. For the fraction $\frac{2}{5}$, multiplying both the numerator and denominator by 3 produces $\frac{6}{15}$. When fractions have like denominators, they are added by adding the numerators and keeping the denominator the same: $\frac{5}{15} + \frac{6}{15} = \frac{11}{15}$.

12. D: Using SOHCAHTOA, tangent is $\frac{y}{x}$ for the special triangles. Since the value needs to be negative one, the angle must be some form of 45 degrees or $\frac{\pi}{4}$. The value is negative in the second and fourth quadrant, so the answer is $\frac{3\pi}{4}$ and $\frac{7\pi}{4}$.

13. C: The numerator in the sequence $\left\{\frac{(-1)^{n+1}}{n^2+5}\right\}_{n=0}^{\infty}$ indicates that the sign of each term changes from term to term. The first term is negative because $n = 0$ and $-1^{n+1} = -1^1 = -1$. Therefore, the second